1/25

ISBN 0 947338 66 7

Axiom
Australia

INTRODUCTION

Until the 1920s, when the automobile began to supersede the horse as the main means of transport on Australia's roads and tracks, horse drawn vehicles were the pride and joy of their owners. Their stately, or workman-like, forms graced the streets of cities and the country tracks.

A major boulevard in a capital city, for example, at the turn of the twentieth century, literally bristled with the variety of vehicles. They were a common feature of everyday life.

There were vehicles used for every conceivable form of transport: carrying families and persons of all types; carting manufactured goods; relaying coffins to grave sites; bringing the bread, meat, milk, vegetables and stores to city and suburban households; conveying royalty to grand occasions; moving fodder, wool, wheat, timber and other rural productions to the great ports with their row upon row of tall ships, sails running from masthead to masthead.

In so many ways, the myriad horse drawn vehicles reflected the history of society in that era. Every timetable planned by families, businesses and governments revolved in some way around the notion of how long it would take to move a horse drawn vehicle from point A to point B, and of how many persons, or of how much weight, that vehicle could safely carry.

At that time, part of every person's life was touched at some point by one of these carts, sulkies, gigs, drays, buggies, waggons, broughams, trams, buckboards, traps, phaetons, or stanhopes.

And then there was the question of where and how these vehicles were to be stored. The wealthier inhabitants of towns and country had their coach houses, farmers had their large sheds for the drays and carts, and traders and tradesmen all had their form of shelter for their vehicles. These buildings, their designs created from the vehicles they housed, were essential elements of the society of that era.

Behind the rich diversity of horse drawn vehicles was a series of crafts – many exquisite in their techniques – which created them: the purchase and preparation of superb quality timbers; the blacksmith's work on tyres, springs and chassis; the wheelwright's tiresome task of creating the spokes and rimming the wheel; the coach painter's delicate brush strokes, rules and decorative art. These crafts might seem to have retreated from the harsh gaze of modern times, but, in reality, they are still practised by those who lovingly restore old vehicles or painstakingly create new ones.

For those in past times who used the vehicles there was always the decision about what to buy, and at what cost. The young trader buying a second hand cart or the wealthy person of commerce purchasing a grand brougham or landau were all at the mercy of the market.

One well-to-do sheep farmer recorded in is diary in 1878, with a sense of horror,

Mr Clark brought up our new Phaeton today, it looks very well and I think will suit me nicely – the step low, a double step in the middle of the trap between the wheels. Price £133. I never thought I should have given so much for anything of the sort.

Not long after the same man purchased an 'Express Waggon' and called it 'a very smart affair – strong and well made, it can carry a ton or 12 persons – lamps, cushions and everything in first rate style'.

The type and quality of vehicle a person drove, then, was a good indicator of the social aspirations that person had. Moreover, vehicles were obviously used as a gauge of social position. By exploring such information we can discern something of the characteristics of our past.

A published work like The Coachbuilder Book of Designs provides the reader with a keen insight into the social and cultural values of a bygone age. It also provides a truly excellent view of the importance of various forms of transport of the era. Again, the book shows the intricacy of the coaches' designs and the wonder of the craft of coachbuilding.

Here is a book which we all can appreciate. It provides a marvellous breadth of information, in word and pictures, for both the social historian and the reader fascinated by the past. Our interest need never flag as we are confronted by the images of these vehicles whose beautiful structures once travelled the highways and byways that we now traverse.

Introduction

IN introducing this Book of Designs to Coachbuilders and Wheelwrights of Australasia it is scarcely necessary to say that most of the designs here published in a classified form have appeared in the *Australasian Coachbuilder and Saddler*. They are now reprinted in response to urgent and persistent calls from subscribers of that journal in all parts of the Colonies. Coachbuilders inform us that they have found the supplements so generally useful as a guide to purchasers that they are in constant demand, but, being unbound, many have been given away, mislaid, or lost, or otherwise rendered unavailable for use. We hope that their republication in this permanent form will make good deficiencies, and greatly enhance their value to the trade.

In addition to the plates we have added tables, giving principal measurements of the vehicles shown, and also some general information collected by us from sources not readily accessible, which we hope will prove serviceable.

With the exception of about twelve plates, the whole of the designs here given are original engravings made for the *Australasian Coachbuilder and Saddler* from drawings by the draftsmen of that journal.

J. E. BISHOP & CO.

Alan Horsnell

The Coachbuilder
BOOK OF DESIGNS

Comprising Engravings of

POPULAR AUSTRALASIAN VEHICLES

with

TABLES OF MEASUREMENTS and OTHER INFORMATION

J. E. BISHOP & CO

PUBLISHERS,

65 MARKET STREET, SYDNEY.

169 QUEEN STREET, MELBOURNE.

1901.

INDEX.

FOUR WHEELED VEHICLES.

	No. of Plate.
Abbott Buggy (Double)	28
Abbott Buggy (Hooded)	32
Abbott Buggy (Open)	46
Angus Buggy	71
Brougham (Single)	3
Brougham	4
Bracket Front Buggy Waggon	77
Buckboards	87, 88
Buggy Waggons	74, 76, 77, 80, 81, 82, 86
Canoe Front Buggy Waggons	82, 86
Canopy Top Traps	26, 90, 92
Coalbox Buggies	37, 42, 45
Concord Buggy	39
Corning Buggies	40, 54
Convertible Seat Buggies	56, 57, 58, 59, 60, 61, 62
Creamery Waggons	153, 154
Dog Carts	19, 21, 22, 24
Delivery Waggon	152
Extension Top Buggy	18
Excelsior Buggy (Single)	48
Excelsior Buggy (Double)	73
Express Adelaide	79, 83
Express Canopy Top	90
Express (Open-sided)	75, 90
Express (Squatters)	78
Express Waggons	89, 91, 93
Franklin Convertible Buggies	57, 58
Farmers' Buggy Waggons	74, 76, 80, 81, 82, 86
Farmers' Waggon	157
Four Passenger Traps	25, 84
Four Passenger Traps (with canopy top)	26, 90, 92
Farmer's Spring Waggon	157
Goddard Buggy	31
Hampden Buggy	92
Jump Seat Buggies	68, 70
Landau	2
Lorry	158
Marni Buggy	43
Market Waggon	155
Milk Waggon	156
Nundah Trap	23
Phaetons	7, 8, 9, 10, 11, 12, 13, 14, 15, 16, 17
Piano Box Buggies	33, 34, 35, 36, 50, 52
Runabout Buggy	85
Slide Seat Hooded	30
Single Buggies	38, 47, 49
Speeding Buggies	53, 55
Station Trap	20
Station Waggon	94
Settler's Waggon	151
Turnout-Seat Buggy (Hooded)	29
Turnout Seat Buggies (Open,)	63, 64, 65, 66, 67, 69
Turnover Seat Buggy	64
Tray Buggy	27
Table Top Waggons	159, 160
Victoria	1
Village Phaeton	13
Waggonette, Farmer's	72
Waggonette, Private	5
Waggonette, Stanhope	6
Whitechapel Buggies	41, 44, 51

TWO WHEELED VEHICLES.

	No. of Plate.
Alexandra Cart	112
Auckland Road Cart	126
Baker's Delivery Cart	149
Bent Shaft Sulkies,	131, 132, 133, 137
Brassey Cart	119
Butcher's Order Cart	147
Cabriolet	114
Cee Spring Gig	127
Cupid Road Cart	130
Cupid Gig	123
Dog Carts	98, 99, 104
Daisy Cart	118
Dairyman's Spring Dray	150
Gigs	113, 123, 125, 127
Gladstone Cart	121
Governess Car	111
King of the Road Sulky	138
Lennox Gig	113
Lindsay Road Cart	136
Milk Delivery Cart	148
Prince George Cart	95
Pagnel Cart	110
Panel Carts, Curved	105, 106, 108, 122.
Phaeton Sulkies	131, 132, 133, 137
Polo Carts	97, 107
Polo Sulky	138
Pony Cart	105, 107
Ralli Cart	96
Road Carts	115, 116, 120, 121, 126, 128, 129, 130, 136, 141, 142
Rustic Carts	100, 101, 102, 103
Speeding Cart	128
Spindle Seat Cart	117
Spring Cart	109
Southland Road Cart	115
Sulkies	131, 132, 133, 134, 135, 137, 138, 139, 140, 143, 144, 145, 146.
Straight Shaft Sulkies	134, 139, 140, 143, 144, 145, 146
Tauiwha Cart	124
Taraniki Gig	125
Tray Sulkies	135, 144
Victoria Road Cart	120

INDEX.—(Cont.)

MISCELLANEOUS PHOTO ENGRAVINGS.

	No. of Engraving.
Abbott, (Double) ..	165
Ambulance Van ..	177
Baker's Cart ..	185
Band Coach ..	173
Bow Waggon ..	174
Canoe Landau and Pair ..	167
Coach, Thoroughbrace ..	171
Commercial Traveller's Waggon ..	172
Drag, Eighteen Passenger ..	180
Dairyman's Cart, Covered ..	183
Dairyman's Cart ..	184
Dairyman's Delivery Waggon ..	186
Dog Cart, Four Wheel ..	166
Extension Top Phaeton ..	164
Fire Brigade Hose Cart ..	176
Fire Brigade Hose Waggon ..	175
Fruit Grower's Waggon ..	182
Hearse ..	178
Jaunting Car ..	170
Lord Brassey's State Coach ..	162
Landaulet ..	163
Popular Turnout, A ..	168
Pony Turnout, A ..	169
Palace Car ..	179
Queen of Holland's State Carriage ..	161
Road Cart ..	181

GENERAL INFORMATION.

Axles, Estimated Carrying Capacity of ..	193
Beams. Relative Strength of ..	197
,, Factors of Safety ..	198
Curves, to Obtain Natural ..	193
Circles, Circumference of ..	198
Ellipse, to Draw an ..	193
Leading Bars, Length of ..	200
Measurements, Table of ..	185
,, General, of English Carriage Doors	199
Poles, Length of, etc., for Pair Horse Vehicles ..	200
Springs, Strength of ..	194
,, Rules for Ordering ..	198
Shafts, Length of, etc., for English Two-Wheeled Vehicles ..	200
Steel, Colours in Tempering ..	200
Timbers, Weight, Strength, and Elasticity of ..	196
Timbers, Shrinkage and Loss of Weight, etc. ..	200
Tyre Steel and Round Edge Iron, Approximate Weight of ..	198

No. 1. VICTORIA.

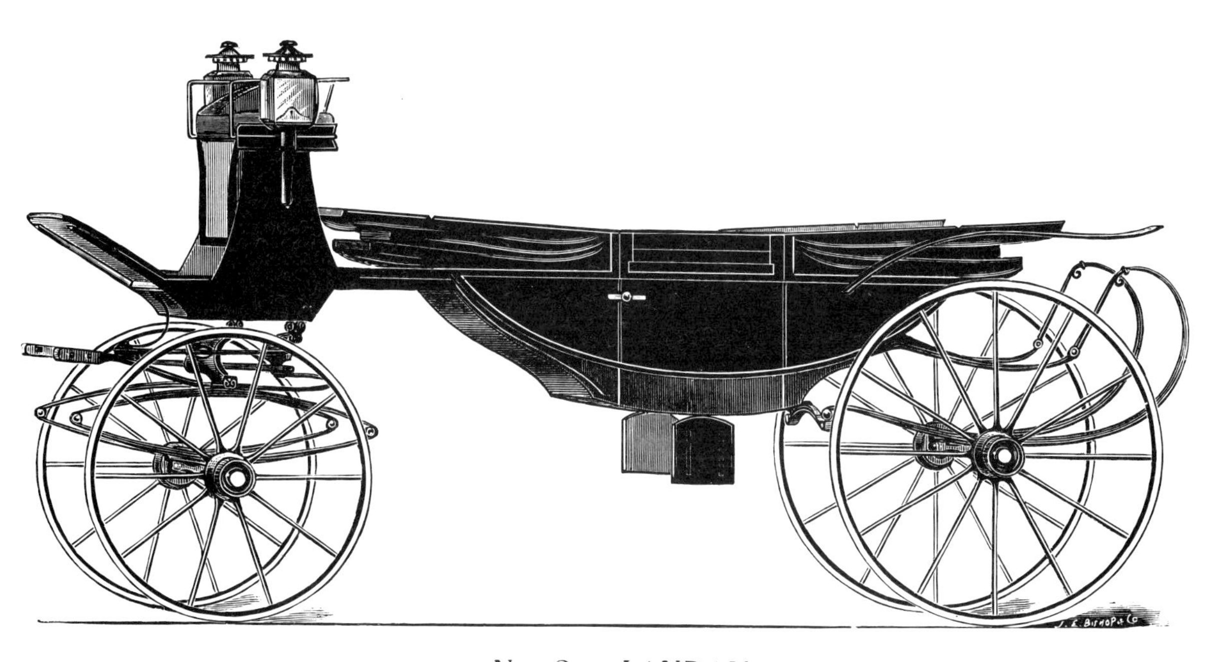

No. 2. LANDAU.

For Sizes, see Table of Measurements.

No. 3. SINGLE BROUGHAM.

No. 4. BROUGHAM.

For Sizes, see Table of Measurements.

No. 5. PRIVATE WAGGONETTE.

No. 6. STANHOPE WAGGONETTE.

For Sizes, see Table of Measurements.

No. 7. OPEN PONY PHAETON.

No. 8. HOODED PONY PHAETON.

ENGLISH FORECARRIAGE.

For Sizes, see Table of Measurements.

No. 9. HOODED PONY-PHAETON.

No. 10. HOODED PONY-PHAETON.

For Sizes, see Table of Measurements.

No. 11. OPEN PONY-PHAETON.

No. 12. HOODED PONY-PHAETON.

ENGLISH CARRIAGE.

For Sizes, see Table of Measurements.

No. 13. VILLAGE PHAETON.

No. 14. HOODED PONY-PHAETON.

For Sizes, see Table of Measurements.

No. 15. OPEN DROP FRONT PHAETON.

No. 16. HOODED DROP FRONT PHAETON.

For Sizes, see Table of Measurements.

No. 17. OPEN PHAETON.

No. 18. EXTENSION TOP TWO-SEAT BUGGY.

For Sizes, see Table of Measurements.

No. 19. FOUR-WHEEL DOG CART.

No. 20. STATION TRAP.

For Sizes, see Table of Measurements.

No. 21. CURVED PANEL FOUR-WHEEL DOG CART.

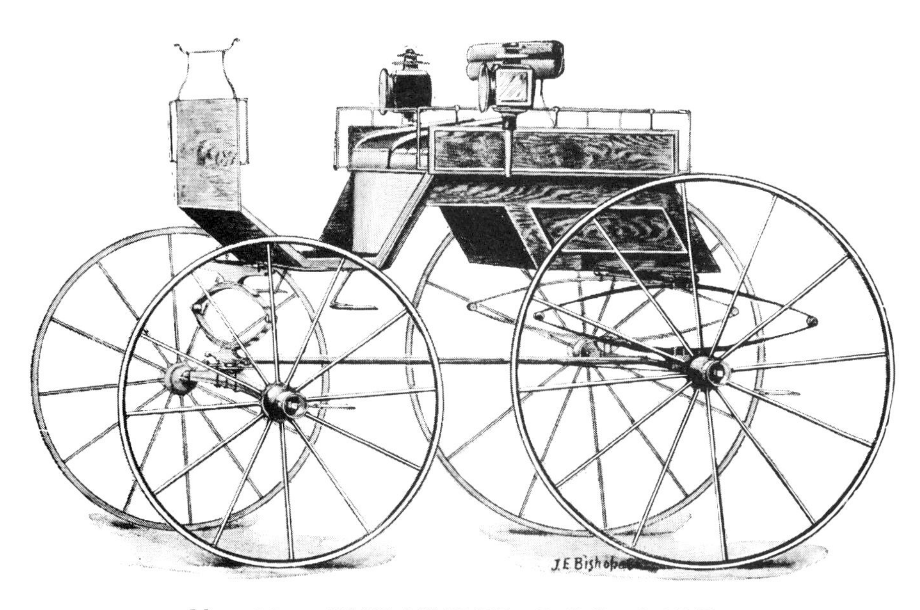

No. 22. CUT-UNDER DOG CART.

For Sizes, see Table of Measurements.

No. 23. NUNDAH TRAP.

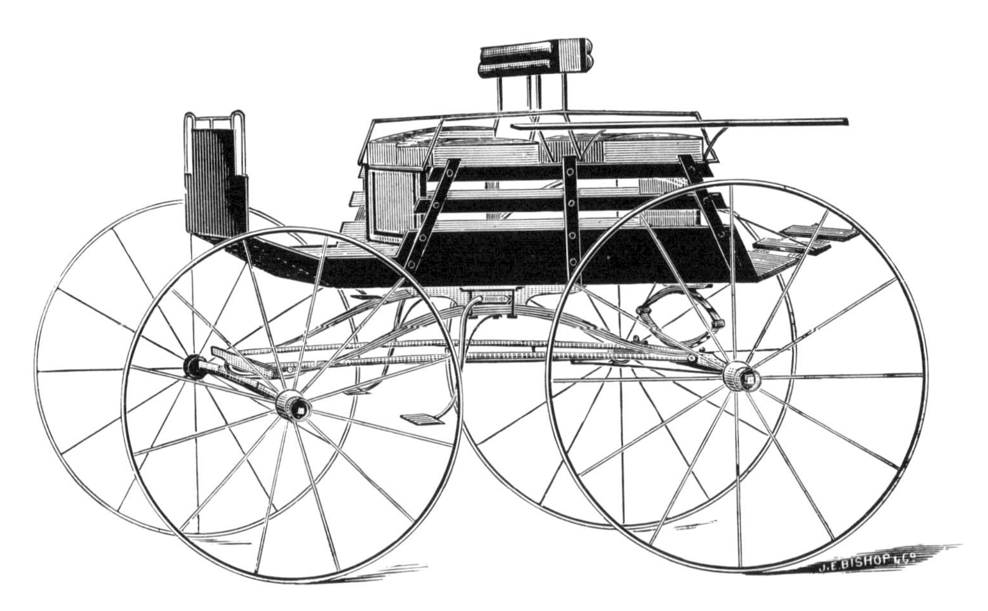

No. 24. SLATTED DOG CART.

ON SIDE SPRINGS.

For Sizes, see Table of Measurements.

No. 25. FOUR PASSENGER TRAP.

No. 26. TRAP WITH CANOPY TOP.

For Sizes, see Table of Measurements.

No. 27. TWO-SEAT TRAY BUGGY.

THE COACHBUILDER BOOK OF DESIGNS.

No. 28. Double Abbott Buggy.

No. 29. HOODED TURN-OUT SEAT BUGGY.

No. 30. HOODED SLIDE-SEAT BUGGY.

For Sizes, see Table of Measurements.

No, 31. GODDARD BUGGY.

No. 32. HOODED ABBOTT BUGGY.

For Sizes, see Table of Measurements.

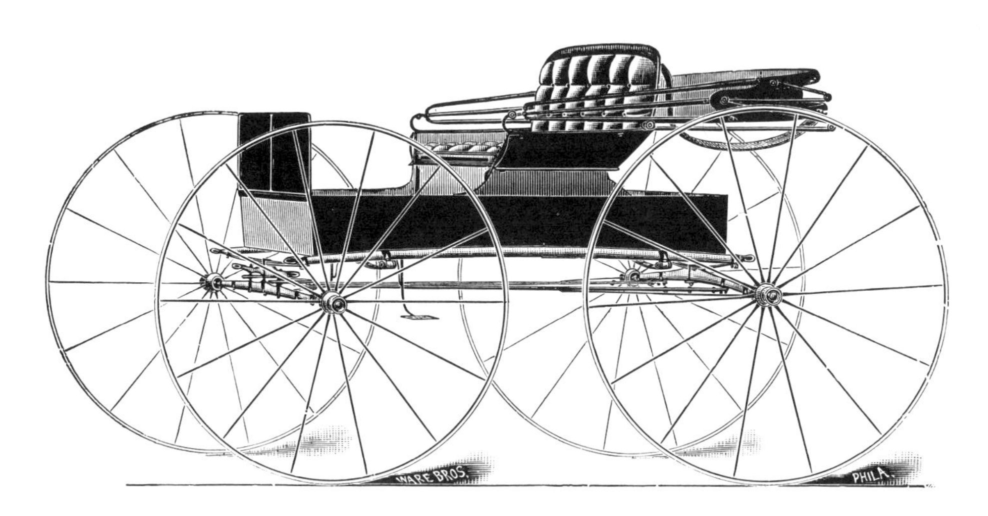

No. 33. PIANO-BOX SIDE-BAR BUGGY.

No. 34. HOODED PIANO-BOX BUGGY.

ON DEXTER UNDER CARRIAGE.

E.

For Sizes, see Table of Measurements.

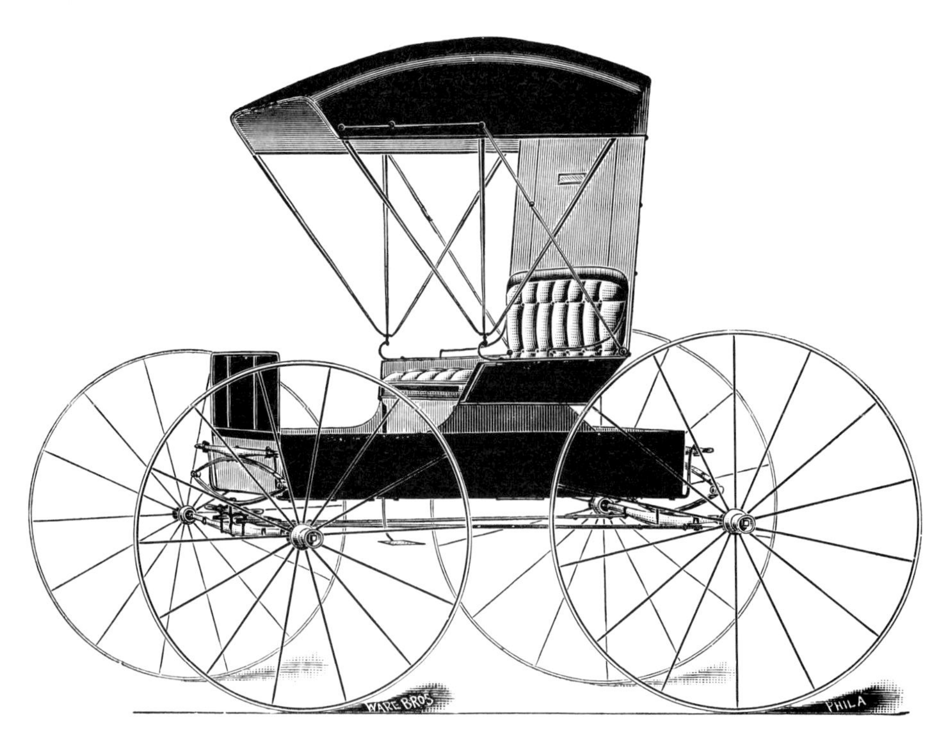

No. 35. HOODED PIANO-BOX BUGGY.

ON ELLIPTIC SPRINGS.

No. 36. HOODED PIANO-BOX BUGGY.

ON SIDE SPRINGS.

For **Sizes**, see Table of Measurements.

No. 37. HOODED COAL-BOX SLIDE-SEAT BUGGY.

SHOWN AS A SINGLE SEAT.

No. 38. HOODED SINGLE BUGGY.

For Sizes, see Table of Measurements.

No. 39. CONCORD BUGGY.

No. 40. HOODED CORNING BUGGY.

ON DEXTER QUEEN UNDER CARRIAGE.

For Sizes, see Table of Measurements.

No. 41. WHITECHAPEL SPEEDING BUGGY.

No. 42. COAL-BOX BUGGY.

For Sizes, see Table of Measurements.

No. 43. MARNI BUGGY.

No. 44. WHITECHAPEL BUGGY.

ON ELLIPTIC SPRINGS.

For Sizes, see Table of Measurements.

No. 45. OPEN COAL-BOX BUGGY.

No. 46. OPEN ABBOTT BUGGY.

For Sizes, see Table of Measurements.

No. 47. SINGLE BUGGY.

No. 48. SINGLE EXCELSIOR BUGGY.

For Sizes, see Table of Measurements.

No. 49. SINGLE SEAT BUGGY.

No. 50. OPEN PIANO-BOX BUGGY.

WITH SPINDLE SEAT.

No. 51. SINGLE WHITECHAPEL BUGGY.

No. 52. OPEN PIANO-BOX BUGGY.

ON SIDE BARS.

For Sizes, see Table of Measurements.

No. 53. SPEEDING BUGGY.

OPEN BACK.

No. 54. CORNING BUGGY.

ON SIDE SPRINGS.

For Sizes, see Table of Measurements.

No. 55. SPEEDING BUGGY.

WITH BOOT.

No. 56. LIGHT CONVERTIBLE-SEAT BUGGY.

For Sizes, see Table of Measurements.

No. 57. FRANKLIN BUGGY.

Shown With One Seat.

No. 58. FRANKLIN BUGGY.

Shown With Two Seats.

For Sizes, see Table of Measurements.

No. 59. CONVERTIBLE BUGGY WAGGON.

SHOWING ONE SEAT.

No. 60. CONVERTIBLE BUGGY WAGGON.

SHOWING TWO SEATS.

For Sizes, see Table of Measurements.

No. 61. CONVERTIBLE SEAT BUGGY WAGGON.

SHOWING ONE SEAT.

No. 62. CONVERTIBLE SEAT BUGGY WAGGON.

SHOWING TWO SEATS.

For Sizes, see Table of Measurements.

No. 63. SHELL-BACK TURN-OUT SEAT BUGGY.

No. 64. TURNOVER SEAT BUGGY.

For Sizes, see Table of Measurements.

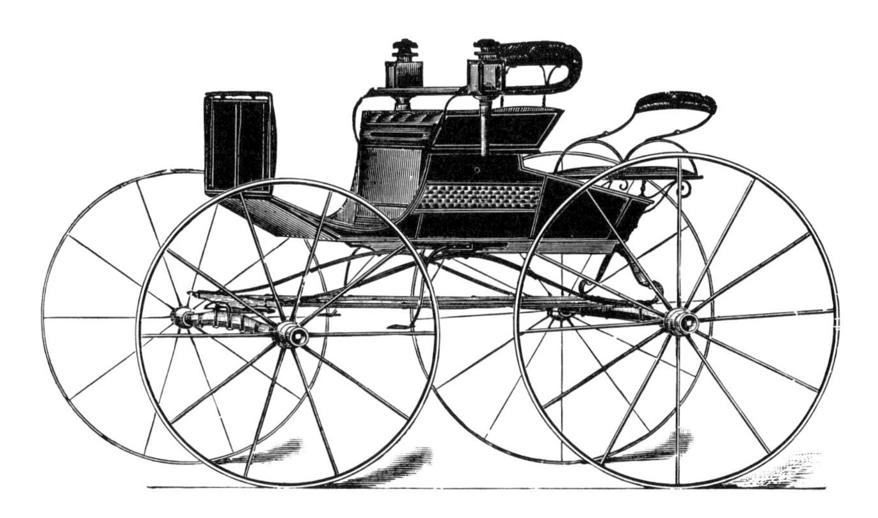

No. 65. TURN-OUT SEAT BUGGY.

No. 66. TURN-OUT SEAT BUGGY.

For Sizes, see Table of Measurements.

No. 67. TURN-OUT SEAT BUGGY.

WITH DROP BRACKET.

No. 68. JUMP SEAT BUGGY.

BACK SEAT ON JUMP IRONS.

For Sizes, see Table of Measurements.

No. 69. TURN-OUT SEAT BUGGY.

COAL-BOX PATTERN.

No. 70. JUMP SEAT BUGGY.

BOTH SEATS ON JUMP IRONS.

For Sizes, see Table of Measurements.

No. **71.** ANGUS BUGGY.

No. **72.** FARMERS' WAGGONETTE.

For Sizes, see Table of Measurements.

No. 73. TWO-SEAT EXCELSIOR BUGGY.

No. 74. FARMER'S BUGGY-WAGGON.

No. 75. TWO-SEAT OPEN-SIDED EXPRESS.

No. 76. FARMER'S TWO-SEAT WAGGON.

For Sizes, see Table of Measurements.

No. 77. BRACKET-FRONT BUGGY-WAGGON.

No. 78. SQUATTER'S EXPRESS.

For Sizes, see Table of Measurements.

No. 79. ADELAIDE EXPRESS.

Both Seats Movable.

No. 80. FARMER'S TWO-SEAT BUGGY-WAGGON.

Back Seat Movable.

For Sizes, see Table of Measurements.

No. 81. FARMERS' BUGGY WAGGON.

BOTH SEATS MOVABLE.

No. 82. CANOE FRONT BUGGY WAGGON.

For Sizes, see Table of Measurements.

No. 83. SOLID SIDED ADELAIDE EXPRESS.

BOTH SEATS MOVABLE. BACK SEAT REVERSIBLE.

No. 84. FOUR PASSENGER TRAP.

For Sizes, see Table of Measurements.

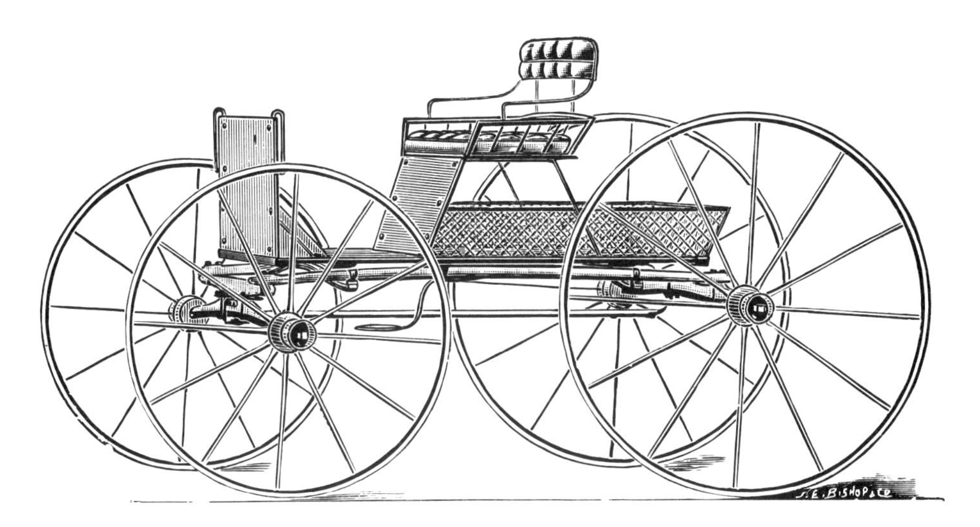

No. 85. RUN-ABOUT BUGGY.

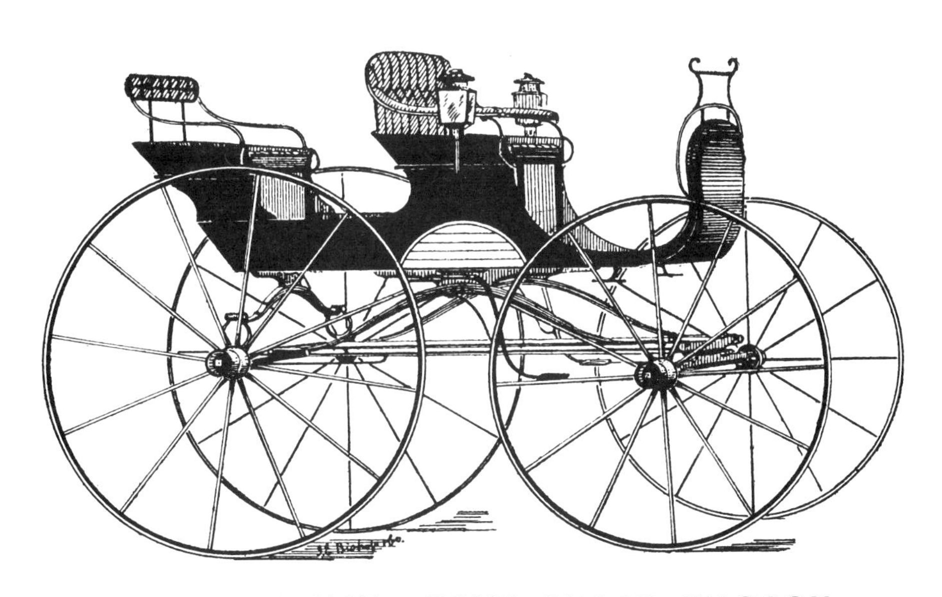

No. 86. CANOE FRONT BUGGY WAGGON.

MOVABLE BACK SEAT.

For **Sizes,** see Table of Measurements.

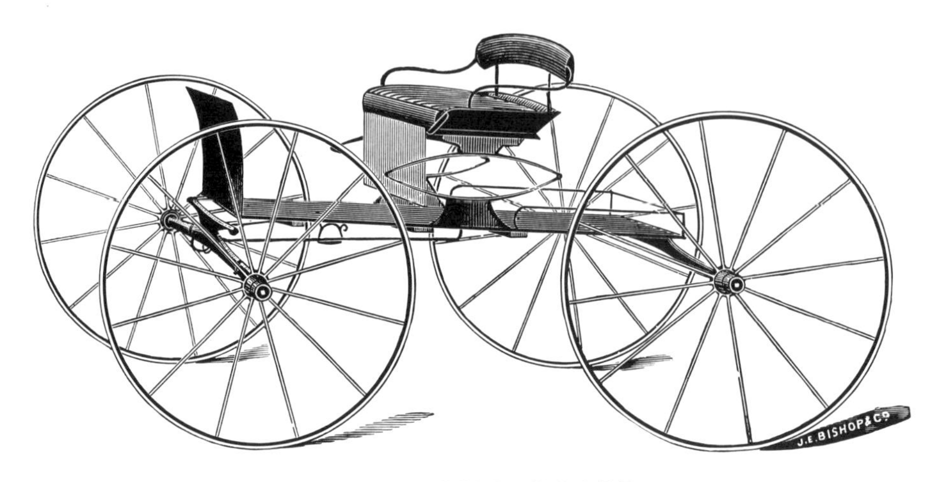

No. 87. BUCKBOARD.

No. 88. QUEENSLAND BUCKBOARD.

For Sizes, see Table of Measurements.

No. 89. EXPRESS WAGGON.

No. 90. OPEN SIDED EXPRESS.

WITH CANOPY TOP.

For Sizes, see Table of Measurements.

No. 91. EXPRESS WAGGON.

WITH TWO MOVABLE SEATS.

No. 92. HAMPDEN BUGGY.

For Sizes, see Table of Measurements.

No. 93. EXPRESS WAGGON.

No. 94. STATION WAGGON.

For Sizes, see Table of Measurements.

No. 95. PRINCE GEORGE CART.

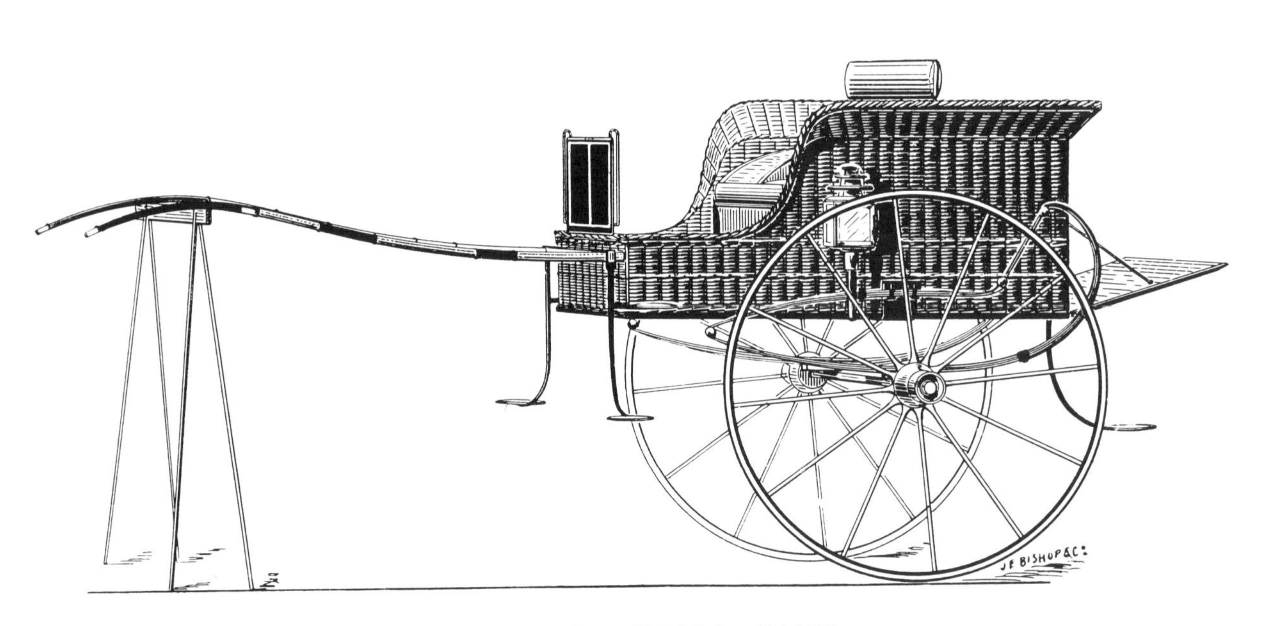

No. 96. RALLI CART.

For Sizes, see Table of Measurements.

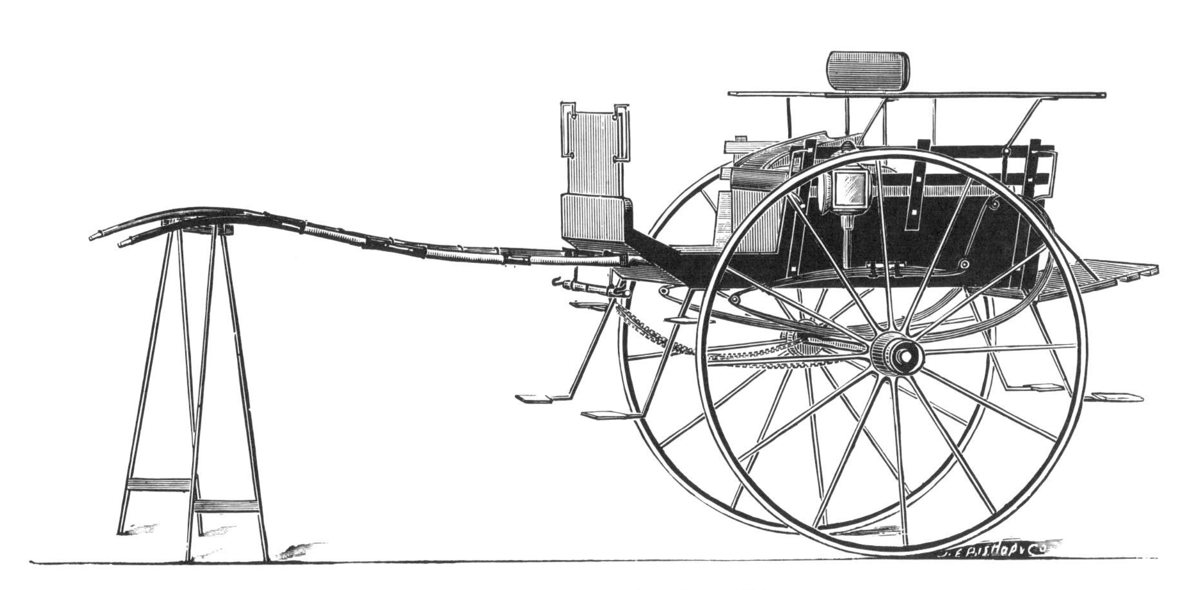

No. 97. POLO CART.

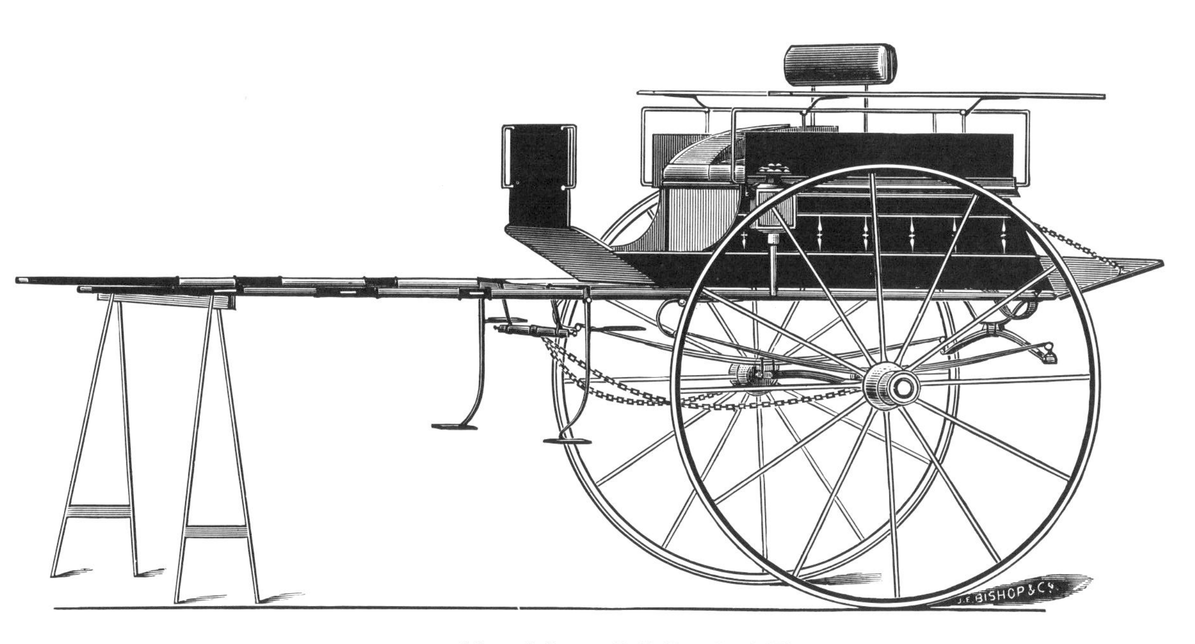

No. 98. DOG CART.

No. 99. DOG CART.

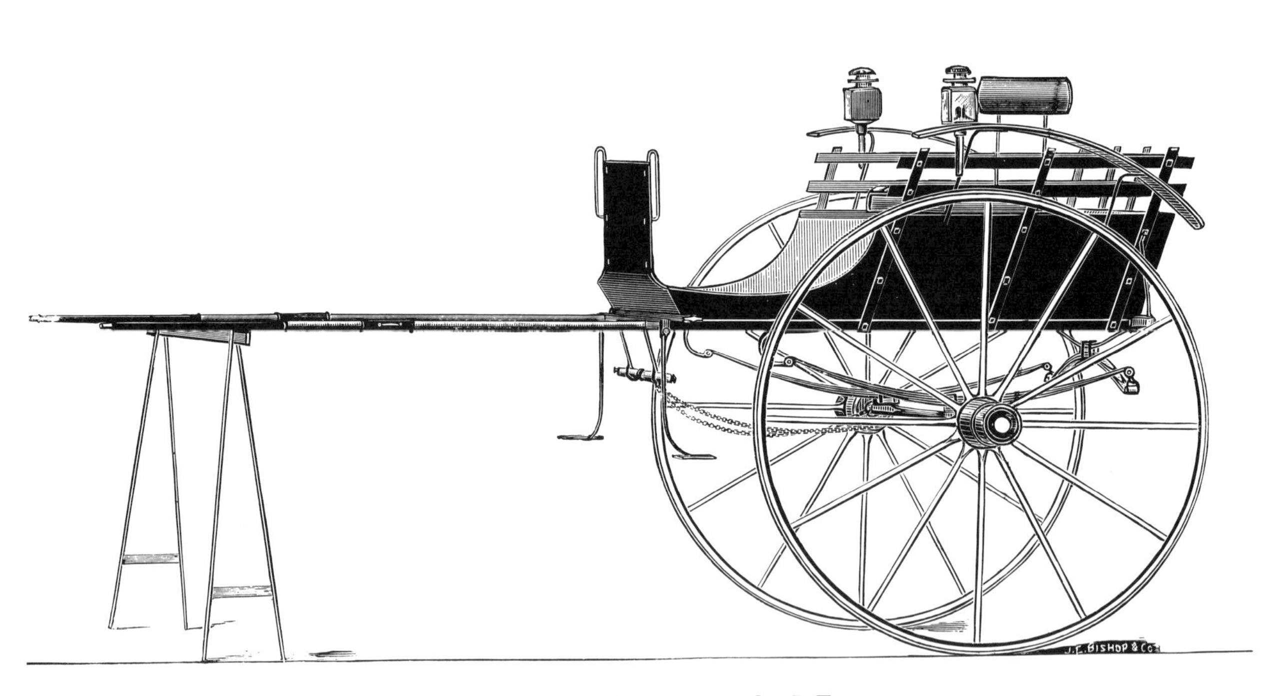

No. 100. RUSTIC CART.

For Sizes, see Table of Measurements.

No. 101. RUSTIC CART.

No. 102. CURVED SIDED RUSTIC CART.

For Sizes, see Table of Measurements.

No. 103. RUSTIC CART.

No. 104. DOG CART.

For Sizes, see Table of Measurements.

No. 105. CURVED PANEL PONY CART.

No. 106. CURVED PANEL CART.

ON CEE SPRINGS.

For Sizes, see Table of Measurements.

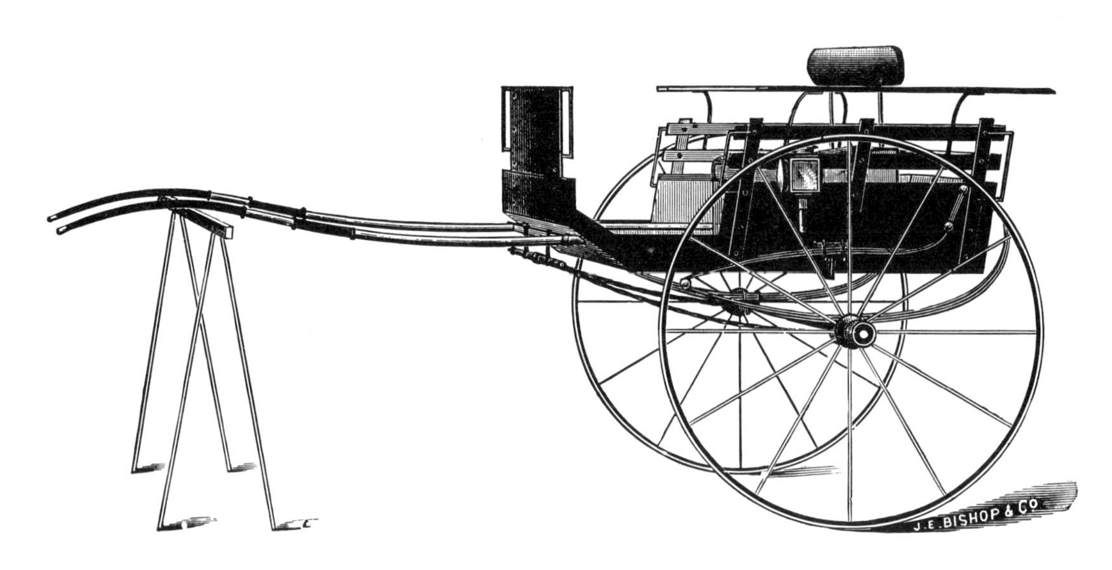

No. 107. PONY POLO CART.

No. 108. CURVED PANEL CART.

For Sizes, see Table of Measurements.

No. 109. SPRING CART.

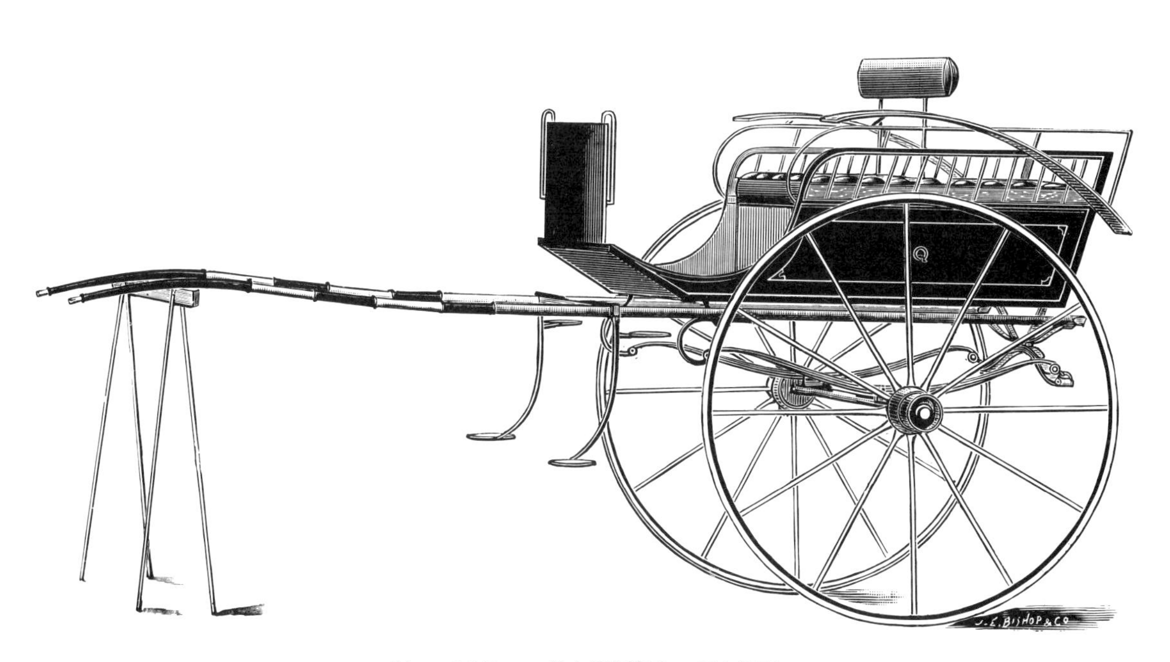

No. 110. PAGNEL CART.

For Sizes, see Table of Measurements.

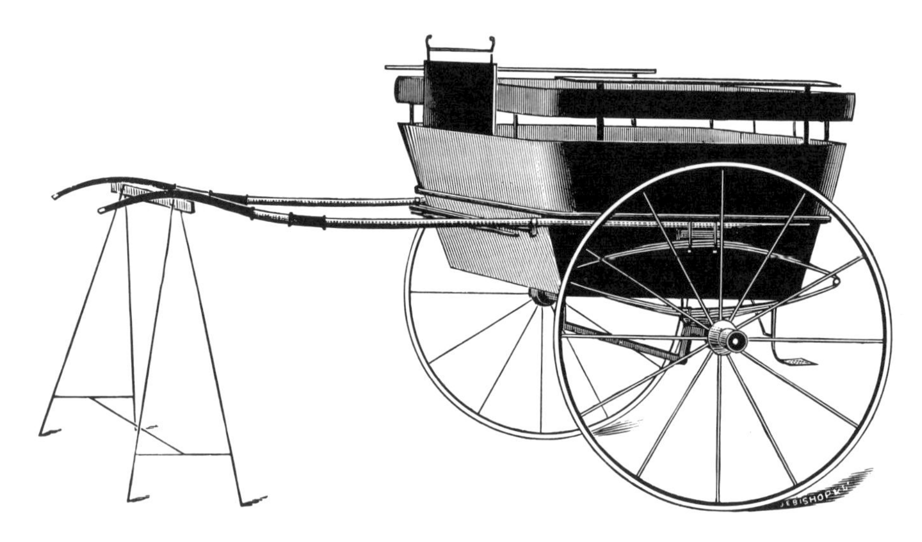

No. 111. GOVERNESS CAR.

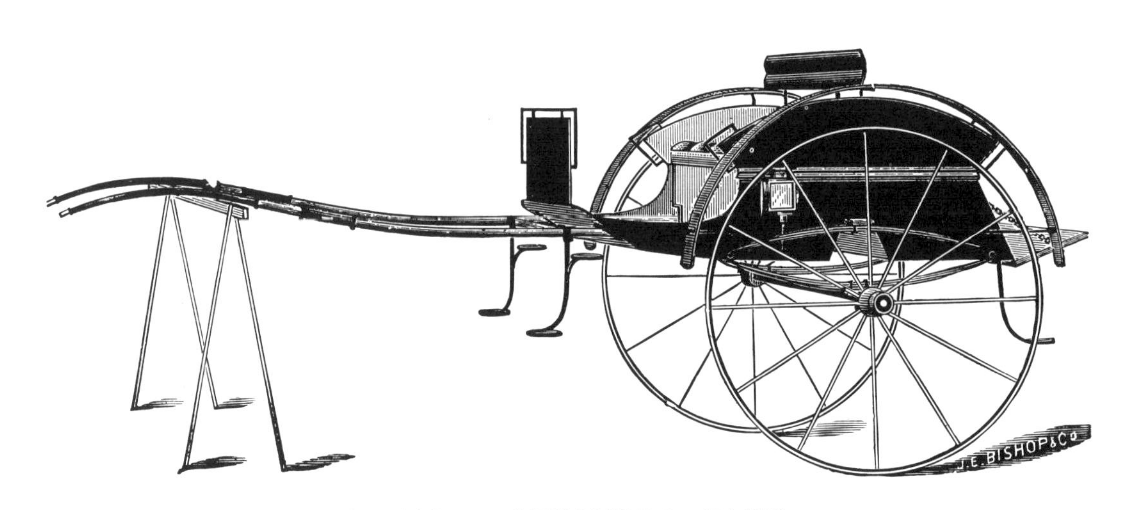

No. 112. ALEXANDRA CART.

For Sizes, see Table of Measurements.

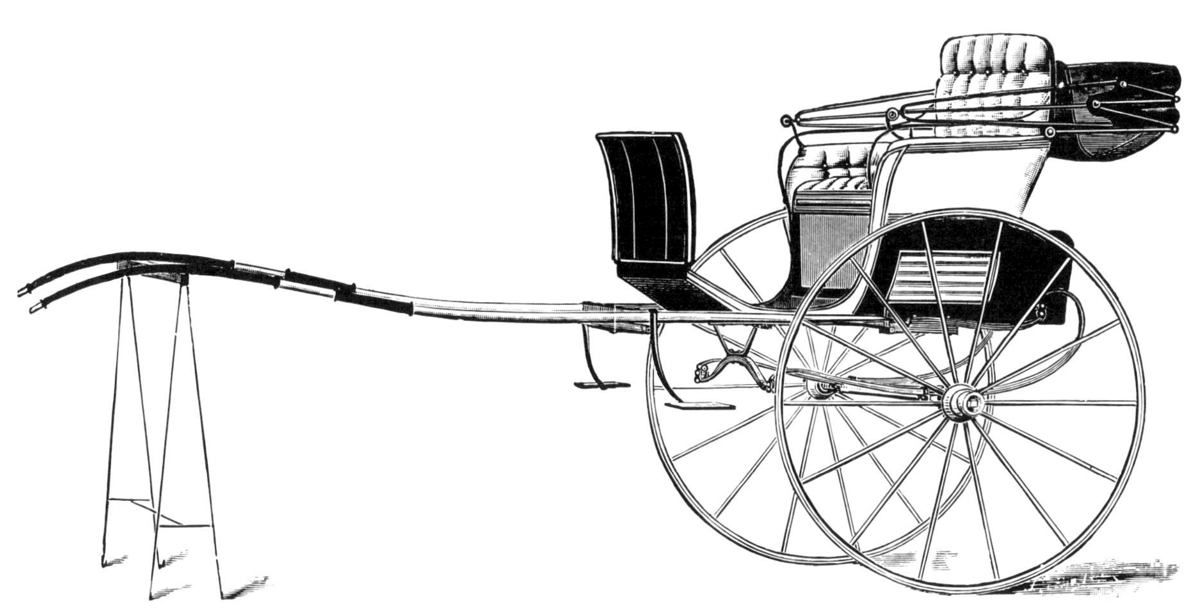

No. 113. LENNOX GIG.

No. 114. CABRIOLET.

For Sizes, see Table of Measurements.

No. **115.** SOUTHLAND ROAD CART.

No. **116.** HOODED ROAD CART.

For Sizes, see Table of Measurements.

No. **117**. SPINDLE SEAT CART.

No. **118**. DAISY CART.

For Sizes, see Table of Measurements.

No. **119**. THE BRASSEY CART.

No. **120**. VICTORIA ROAD CART.

For Sizes, see Table of Measurements.

No. 121. GLADSTONE CART.

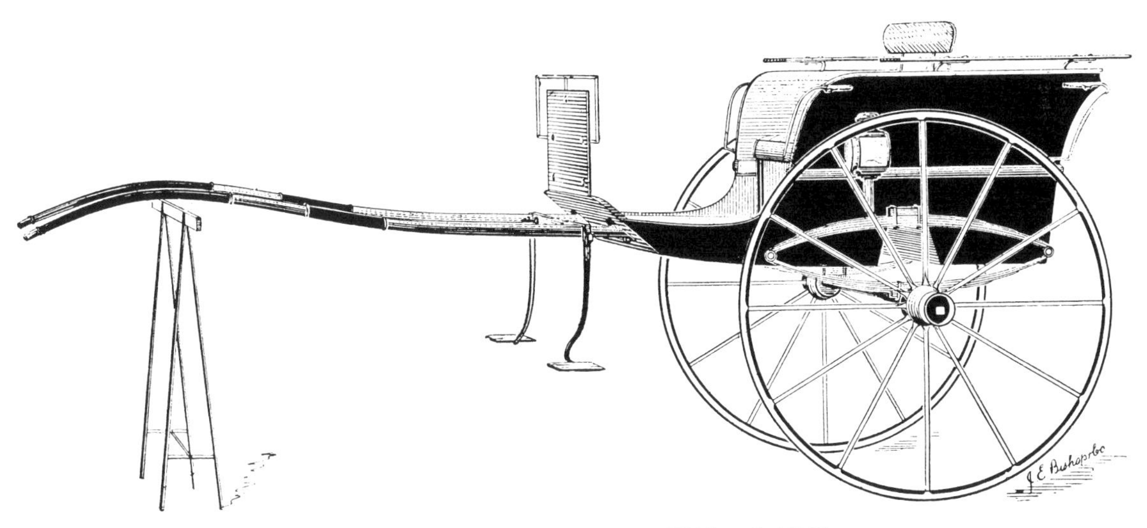

No 122. CURVED PANEL CART.

ON ELLIPTIC SPRINGS

For Sizes, see Table of Measurements.

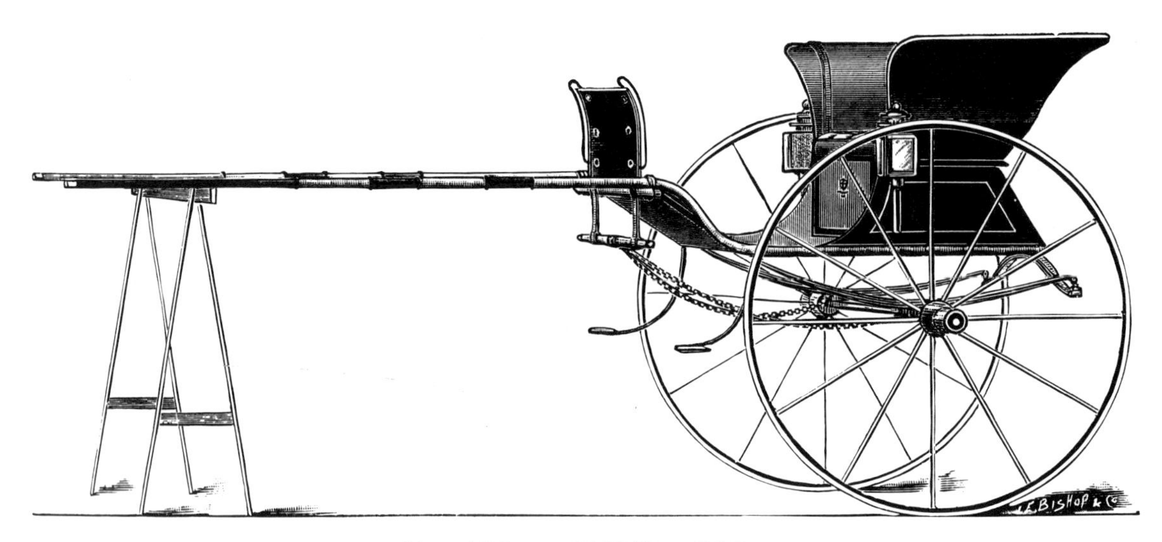

No. **123.** CUPID GIG.

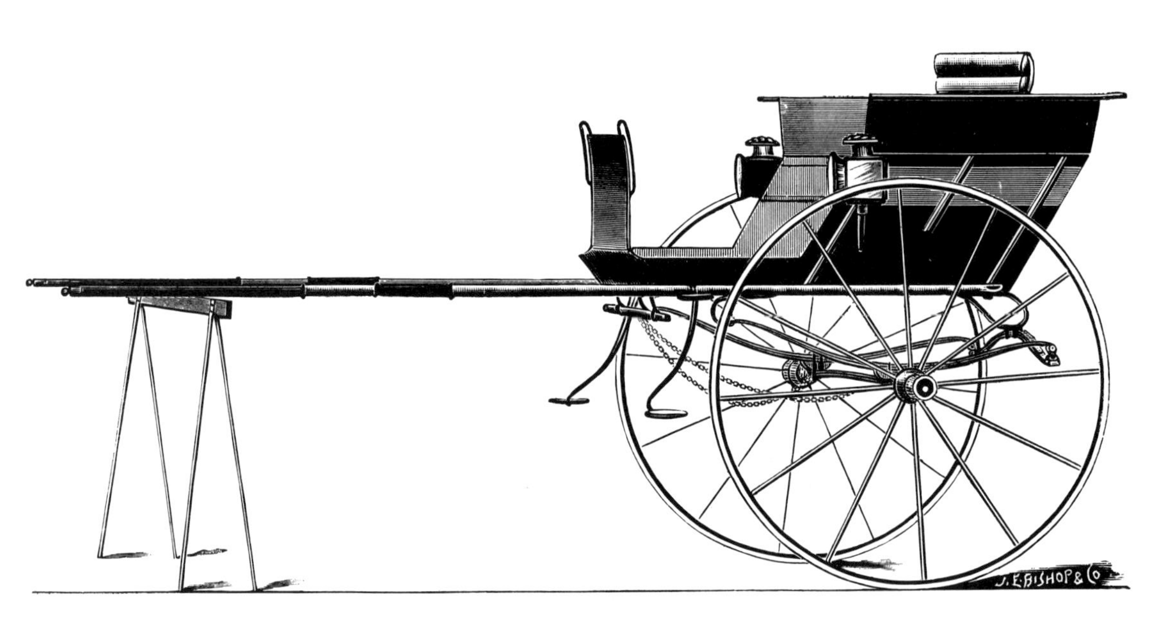

No. **124.** TAUIWHA CART.

For Sizes, see Table of Measurements.

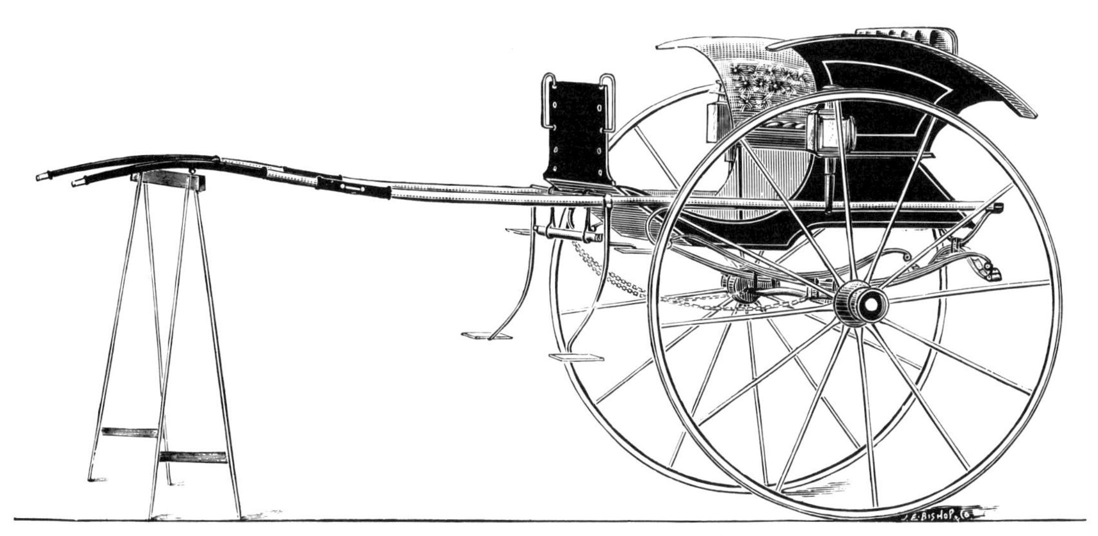

No. 125. TARANAKI GIG.

No. 126. AUCKLAND ROAD CART.

For Sizes, see Table of Measurements.

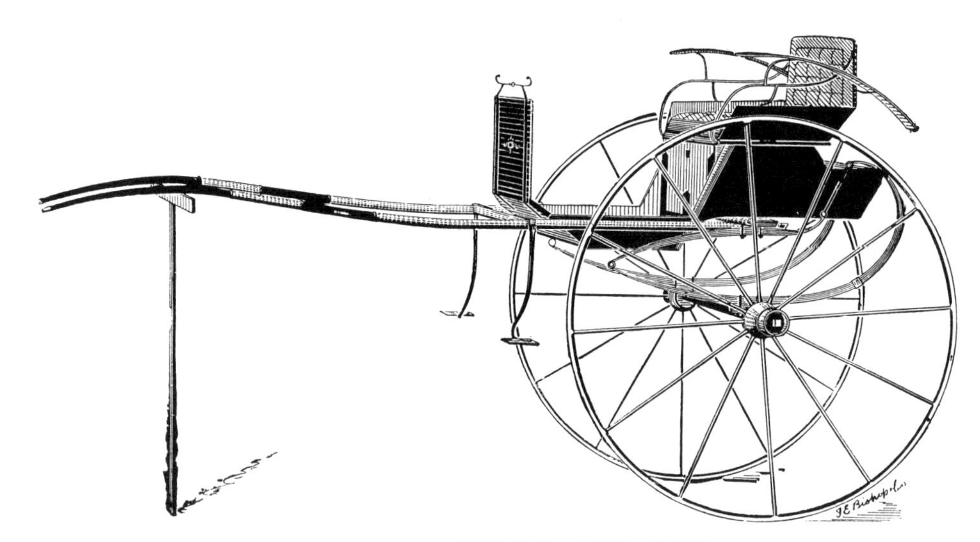

No. **127.** CEE SPRING GIG.

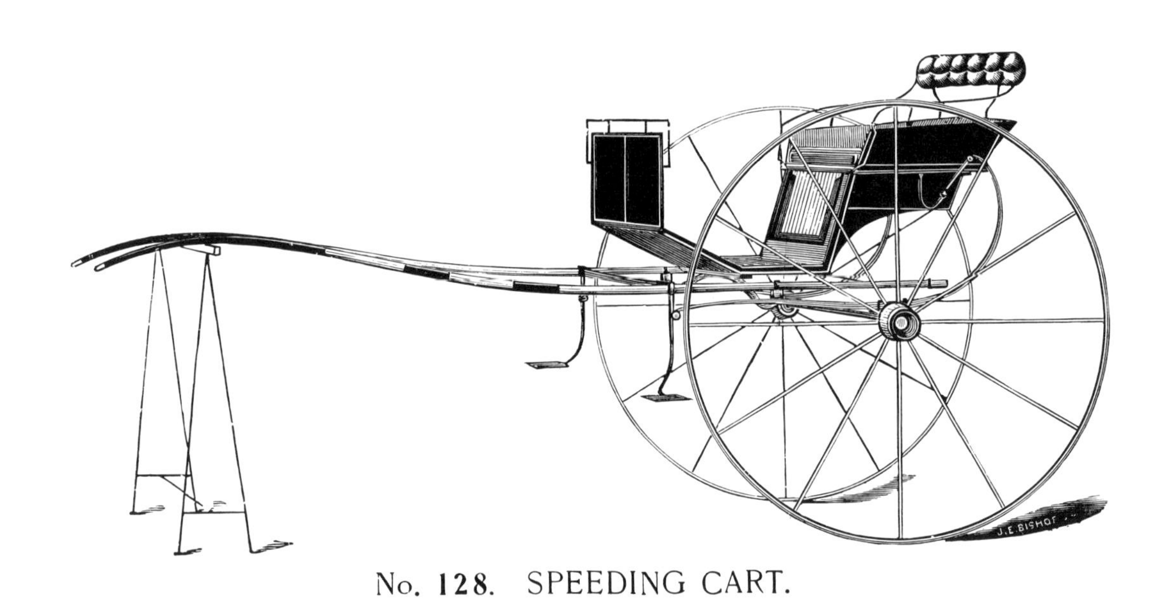

No. **128.** SPEEDING CART.

For Sizes, see Table of Measurements.

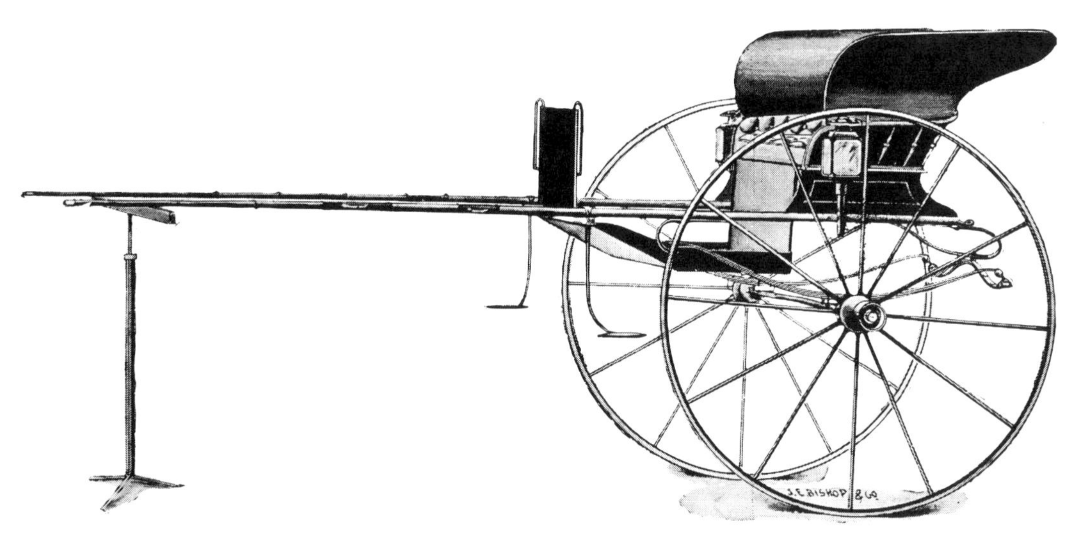

No. **129**. ROAD CART.
WITH CUPID WING BOARDS.

No. **130**. CUPID ROAD CART.

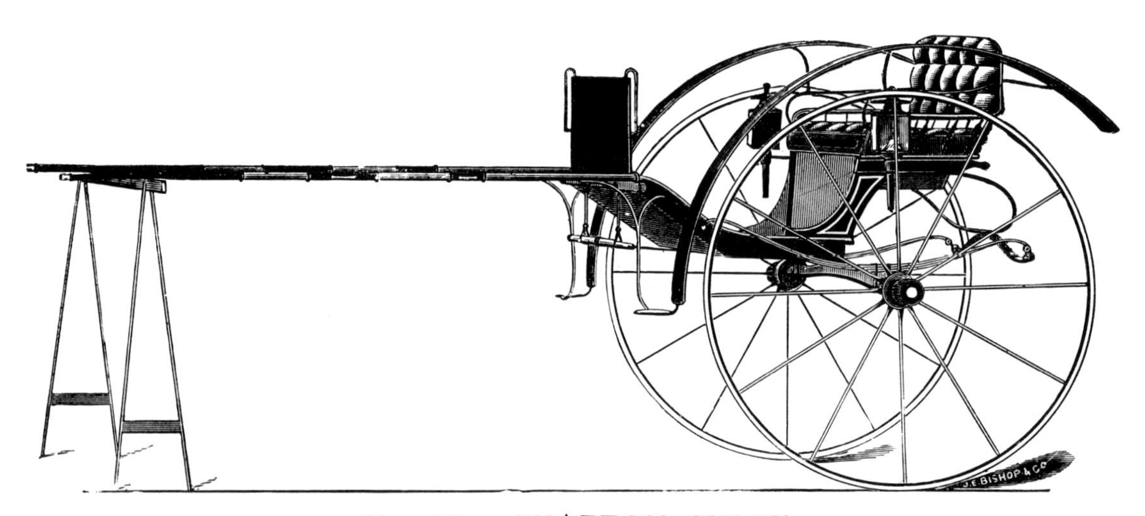

No. 131. PHAETON SULKY.
WITH SPINDLE SEAT.

No. 132. PHAETON SULKY.
WITH SOLID SEAT.

For Sizes, see Table of Measurements.

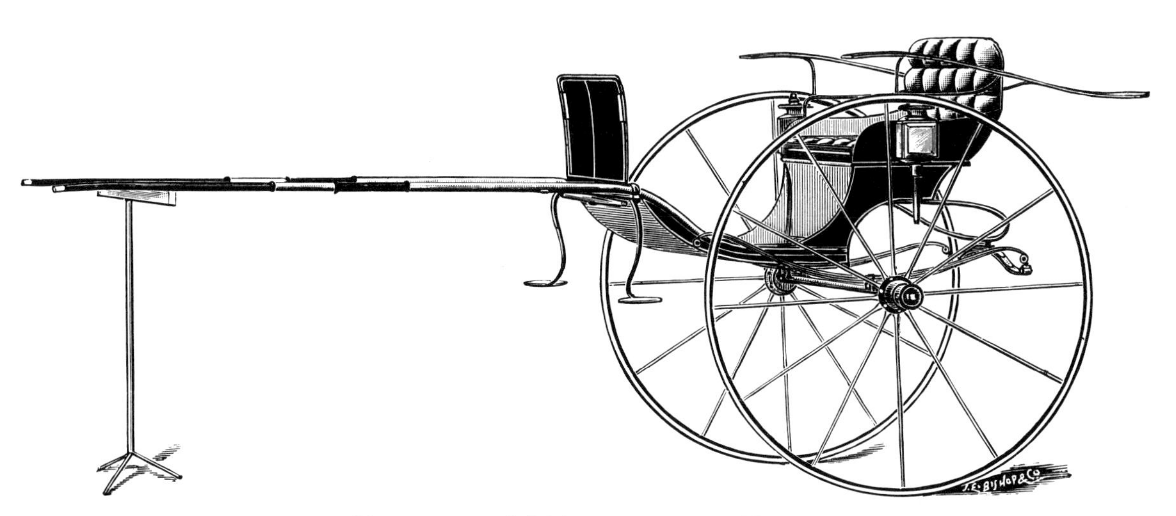

No. **133.** BENT SHAFT SULKY.

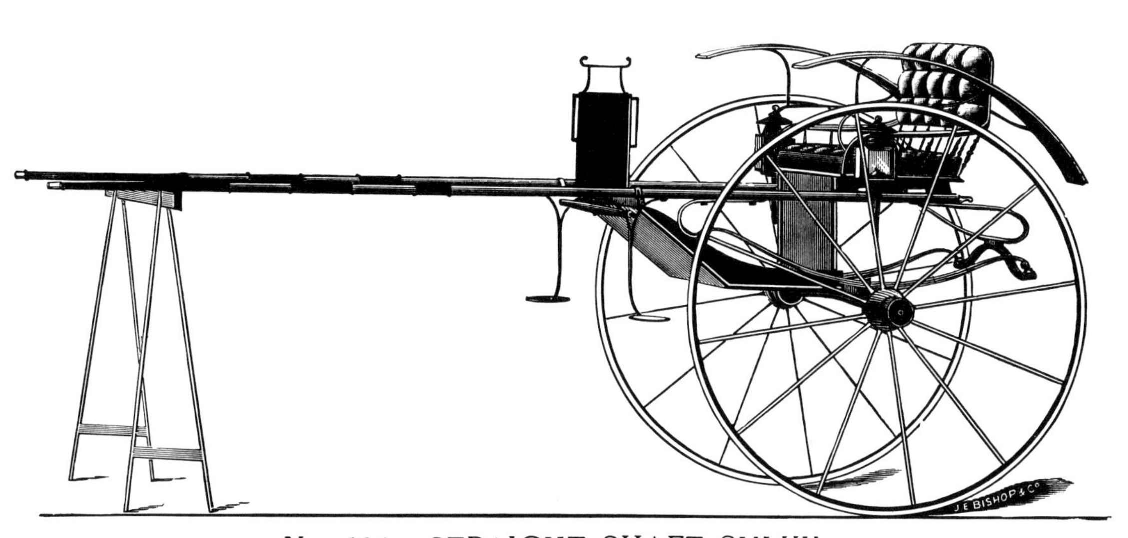

No. **134.** STRAIGHT SHAFT SULKY.

For Sizes, see Table of Measurements.

No. **135.** TRAY SULKY.

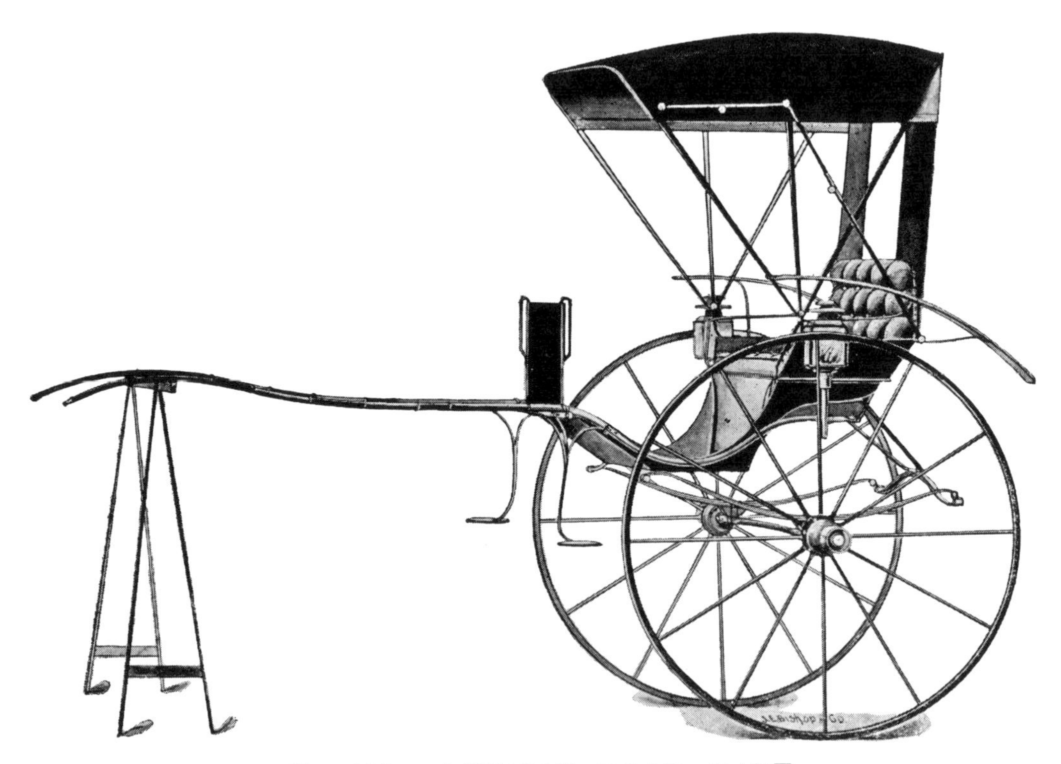

No. **136.** LINDSAY ROAD CART.

For Sizes, see Table of Measurements.

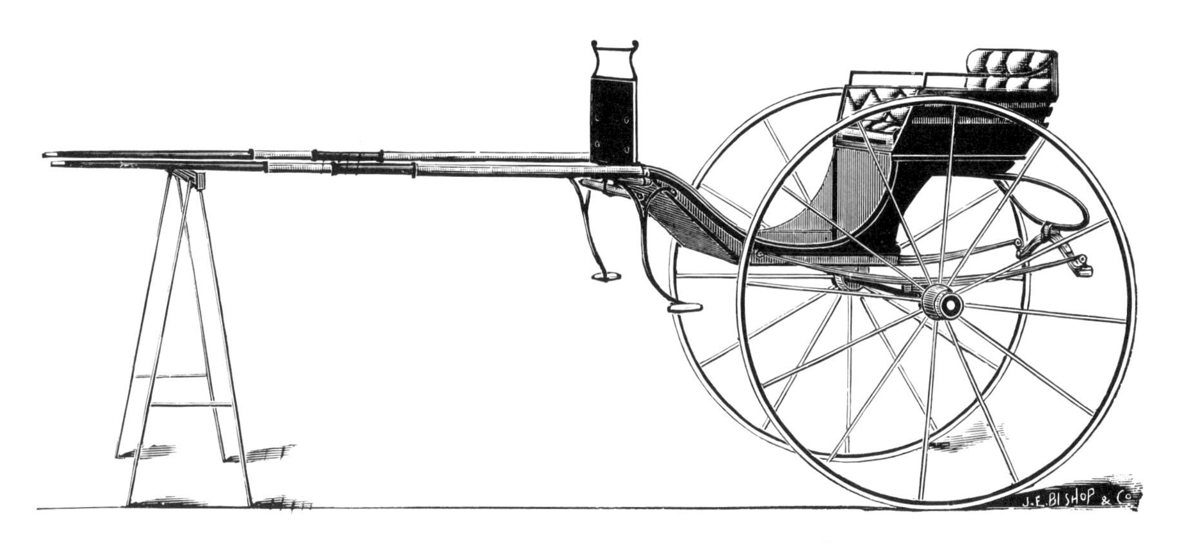

No. 137. BENT SHAFT SULKY.

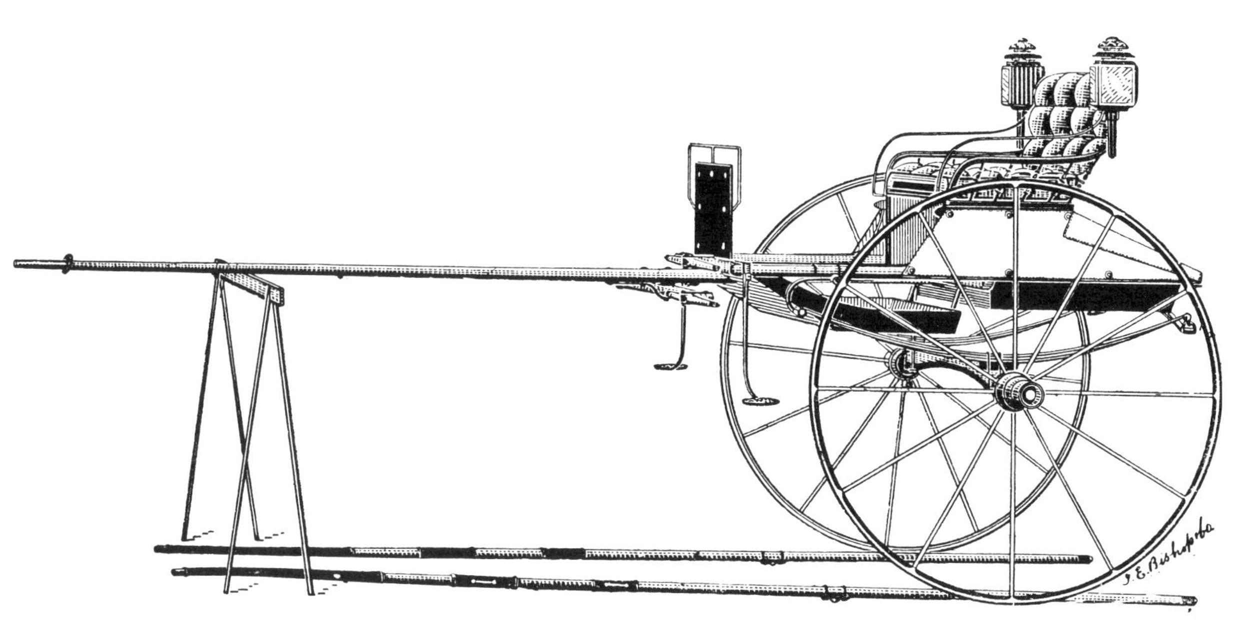

No. 138. KING OF THE ROAD SULKY.
CONVERTIBLE FOR POLE OR SHAFTS.

For Sizes, see Table of Measurements.

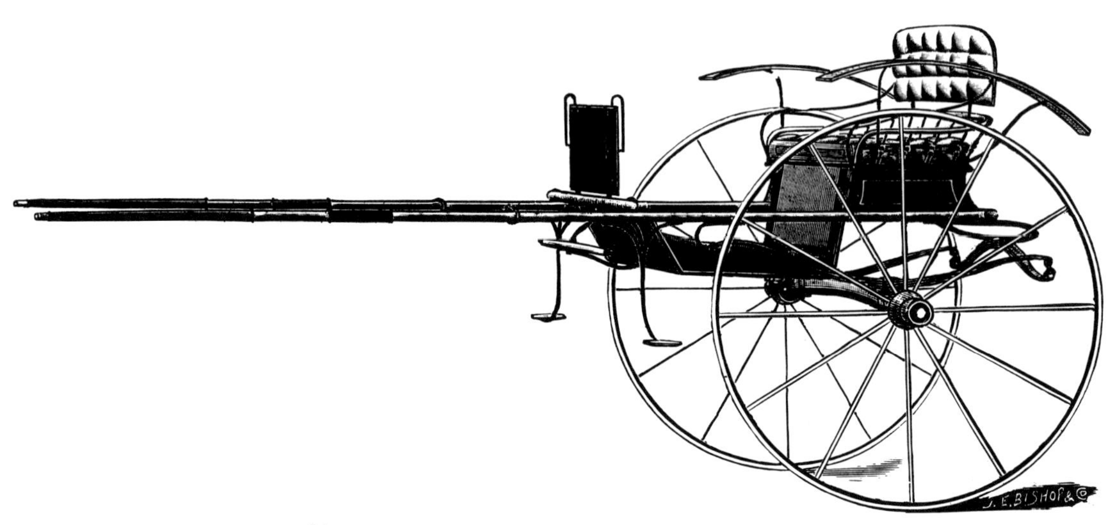

No. 139. SPINDLE SEAT SULKY.

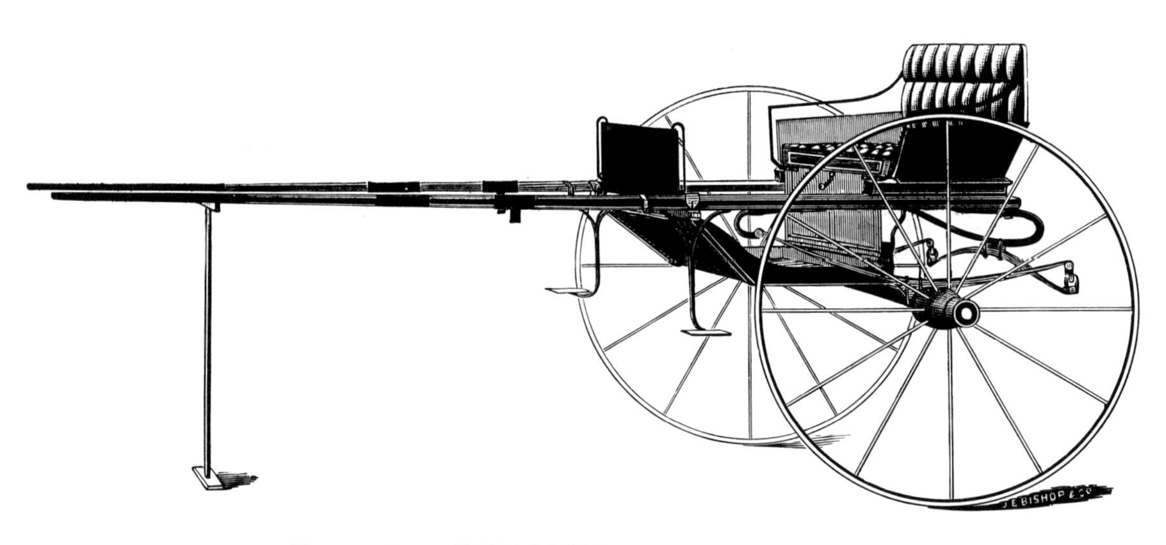

No. 140. STRAIGHT SHAFT SULKY.

WITH SOLID SEAT.

For Sizes, see Table of Measurements.

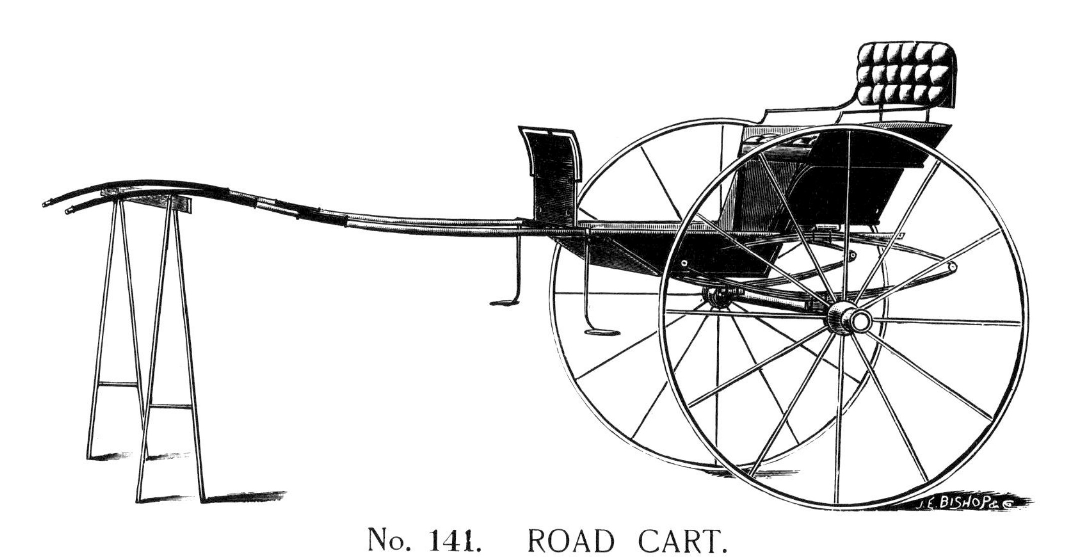

No. 141. ROAD CART.

ON ELLIPTIC SPRINGS.

No. 142. ROAD CART.

ON CEE-ELLIPTIC SPRINGS.

For Sizes, see Table of Measurements.

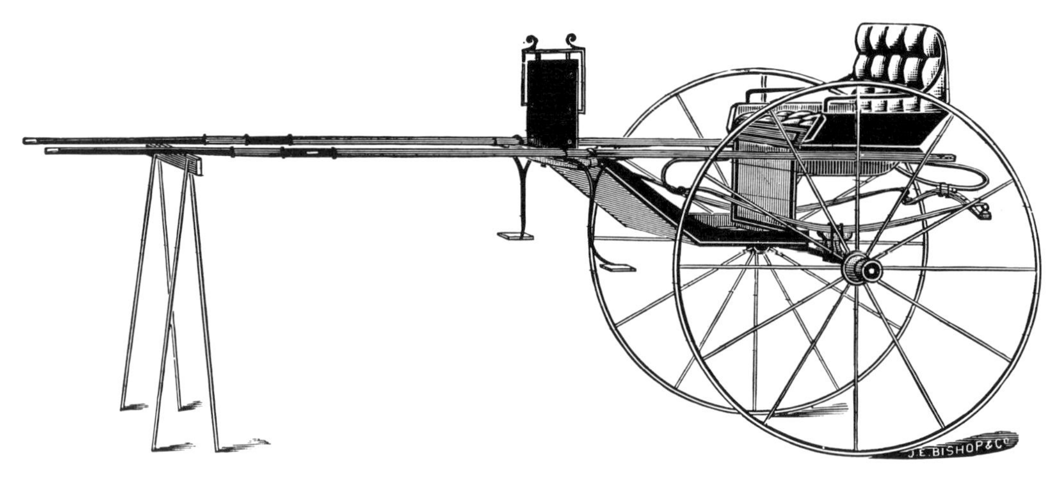

No. 143. STRAIGHT SHAFT SULKY.
WITH SOLID SEAT.

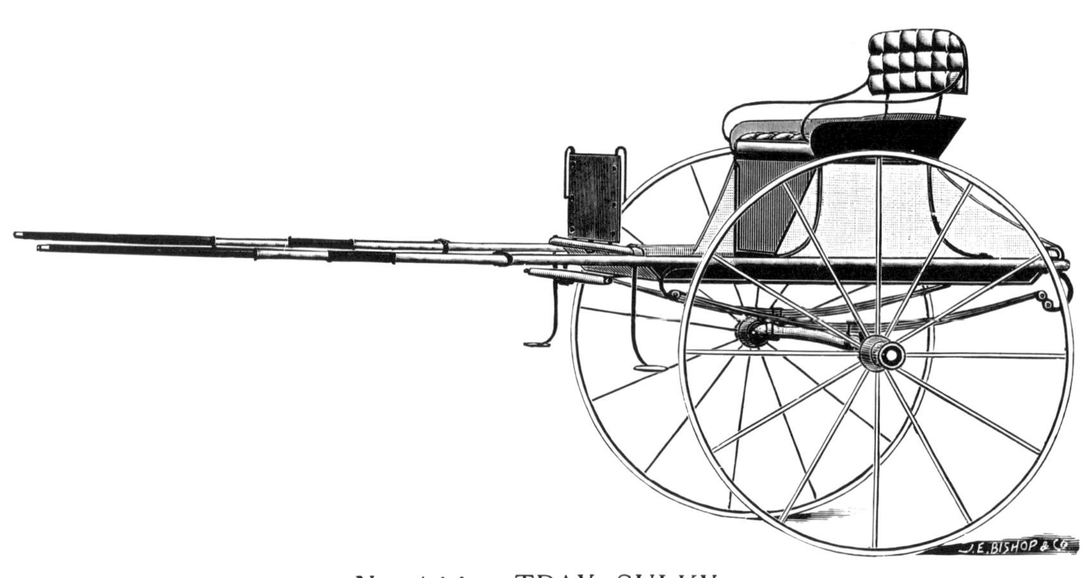

No. 144. TRAY SULKY.
WITH CANVAS BOOT

For Sizes, see Table of Measurements.

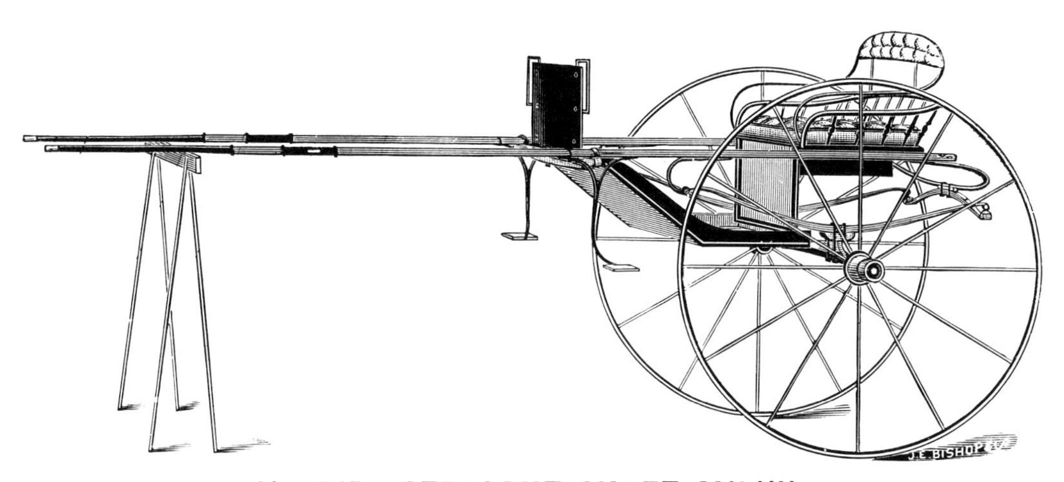

No. **145.** STRAIGHT SHAFT SULKY.
WITH SPINDLE SEAT.

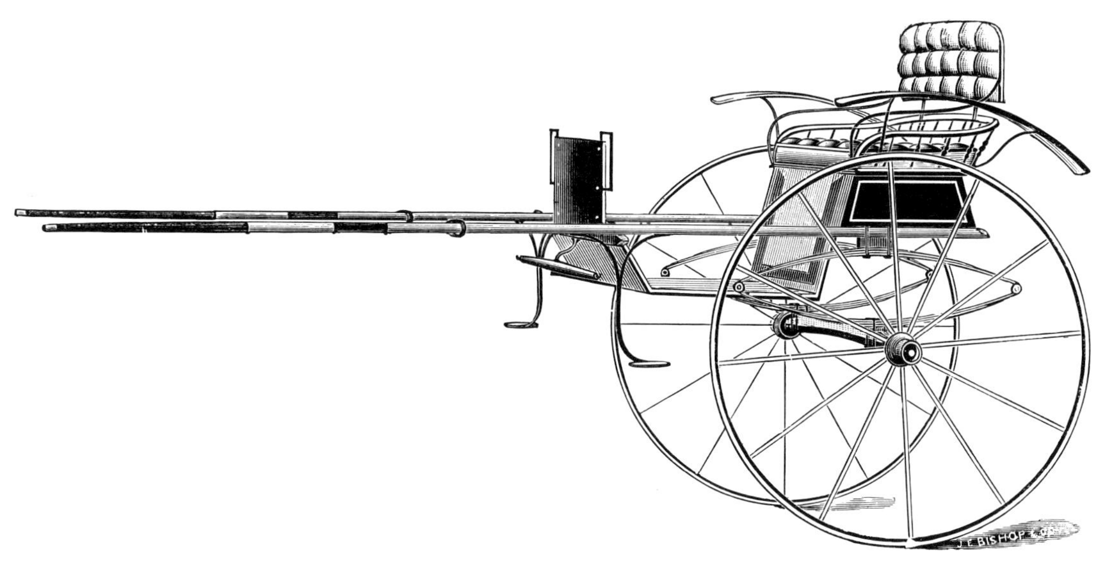

No. **146.** STRAIGHT SHAFT SULKY.
SEAT ON BRACKETS.

For Sizes, see Table of Measurements.

No. **147.** BUTCHER'S ORDER CART.

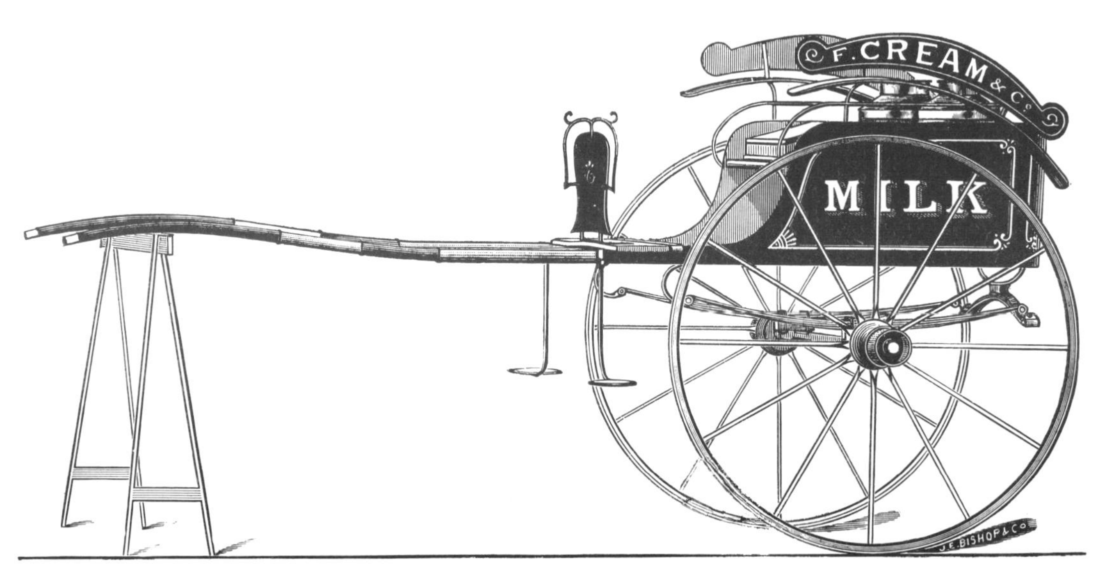

No. **148.** MILK DELIVERY CART.

For Sizes, see Table of Measurements.

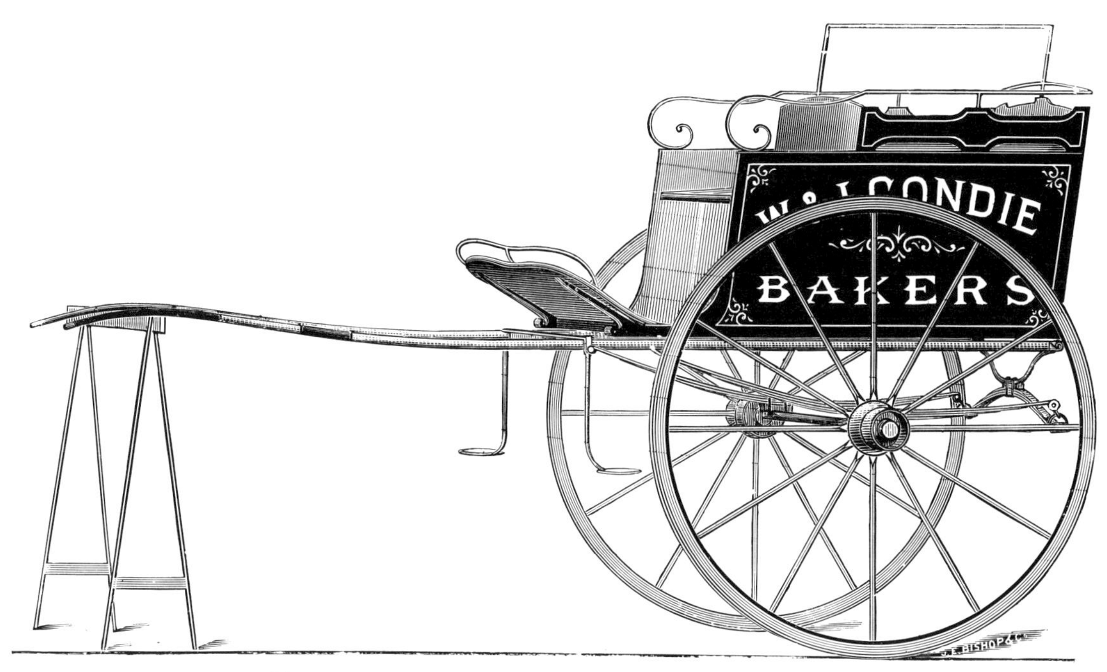

No. **149.** BAKER'S DELIVERY CART.

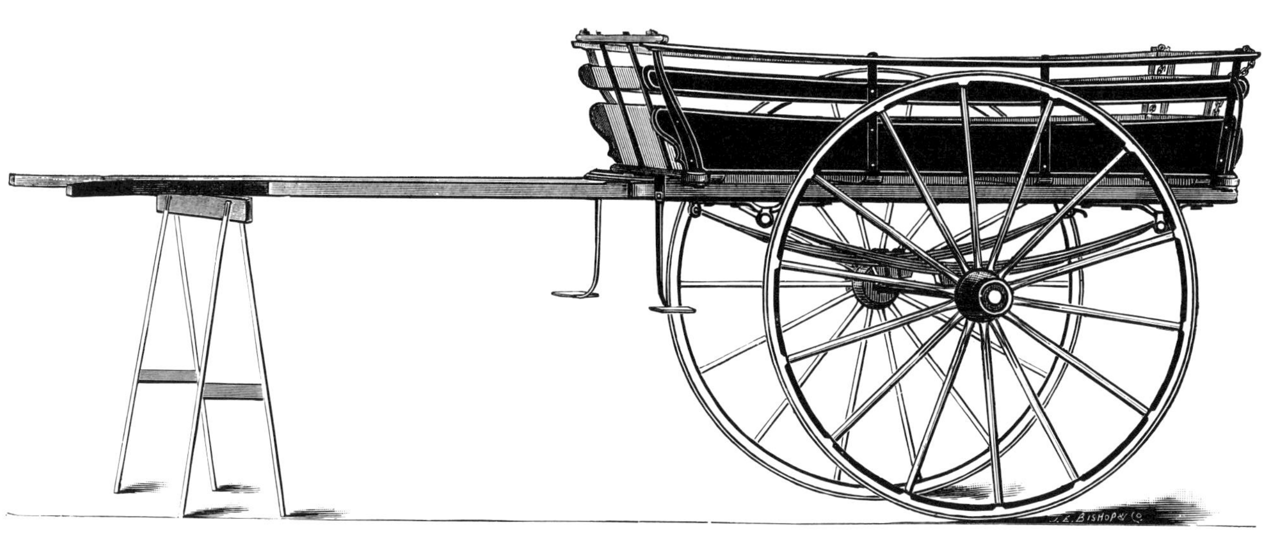

No. **150.** DAIRYMAN'S SPRING DRAY.

For Sizes, see Table of Measurements.

No. **151.** SETTLER'S WAGGON.

No. **152.** BUSINESS DELIVERY WAGGON.

For Sizes, see Table of Measurements.

No. 153. CREAMERY WAGGON.

WITHOUT BATTENS ON SIDES.

No. 154. CREAMERY WAGGON.

WITH BATTENS ON SIDES

For Sizes, see Table of Measurements.

No. 155. MARKET WAGGON.

MOVABLE SIDE BATTENS.

No. 156. MILK WAGGON.

For Sizes, see Table of Measurements.

No. **157.** FARMER'S SPRING WAGGON.

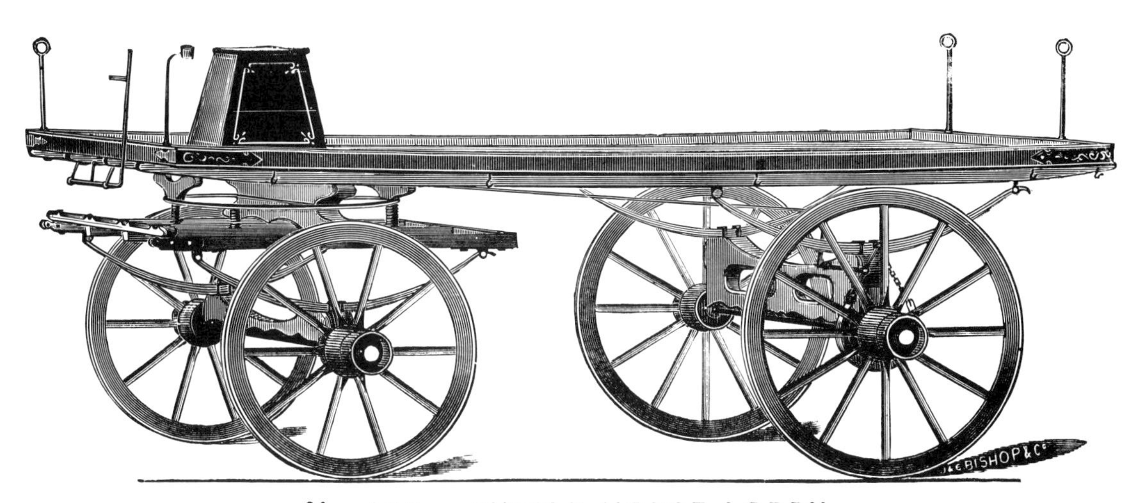

No. **158.** SINGLE HORSE LORRY.

For Sizes, see Table of Measurements.

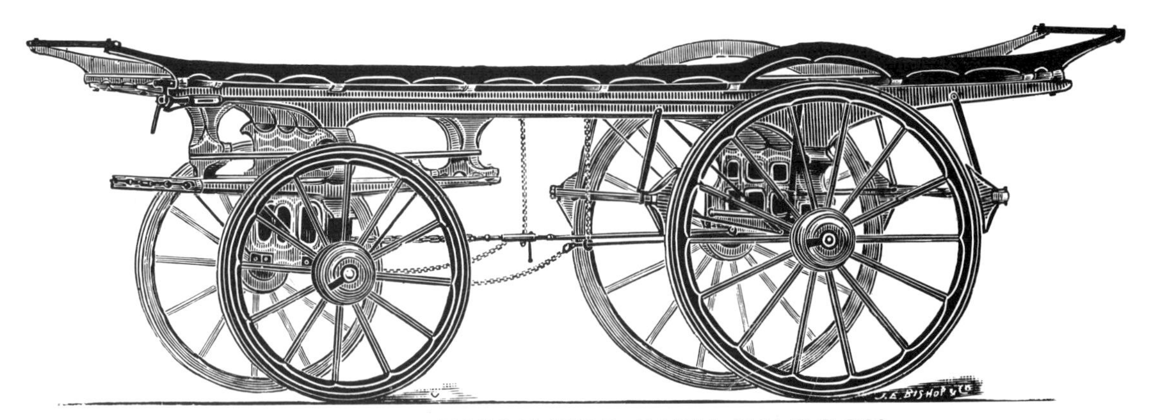

No. **159.** TABLE TOP FARM WAGGON.

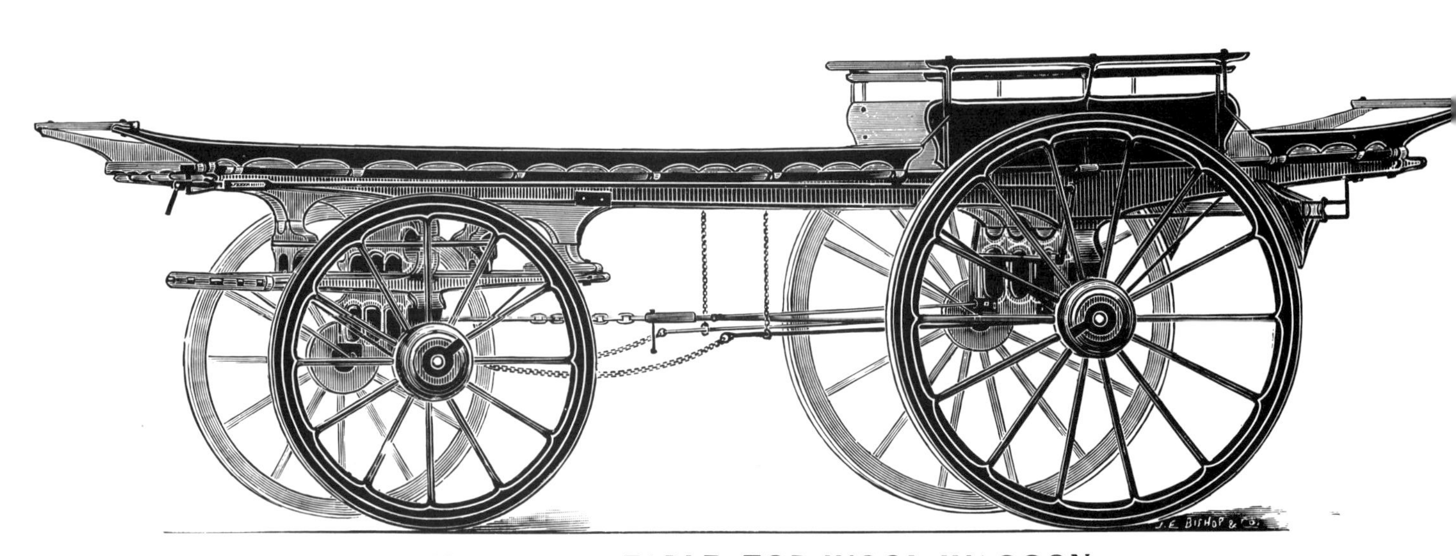

No. **160.** TABLE TOP WOOL WAGGON.

For Sizes, see Table of Measurements.

Photo-Engravings

REPRESENTING

VARIOUS TYPES

OF

VEHICLES

Reprinted from "The Australasian Coachbuilder and Saddler."

No. 161. QUEEN OF HOLLAND'S STATE CARRIAGE.

No. 162. LORD BRASSEY'S STATE CARRIAGE.

No. 163. LANDAULET.

No. 164. EXTENSION-TOP PHAETON.

No. **165**. DOUBLE ABBOTT BUGGY.

No. **166**. FOUR-WHEEL DOG CART.

No. 167. CANOE LANDAU AND PAIR.

No. 168. A POPULAR TURNOUT.

No. **169.** A PONY TURNOUT.

No. **170.** JAUNTING CAR.

No. 171. THOROUGHBRACE COACH.

No. 172. COMMERCIAL TRAVELLER'S WAGGON.

No. **173.** BAND COACH.

No. **174.** BOW WAGGON.

No. **175.** FIRE BRIGADE HOSE WAGGON.

No. **176.** FIRE BRIGADE HOSE CART

No. 177. AMBULANCE VAN.

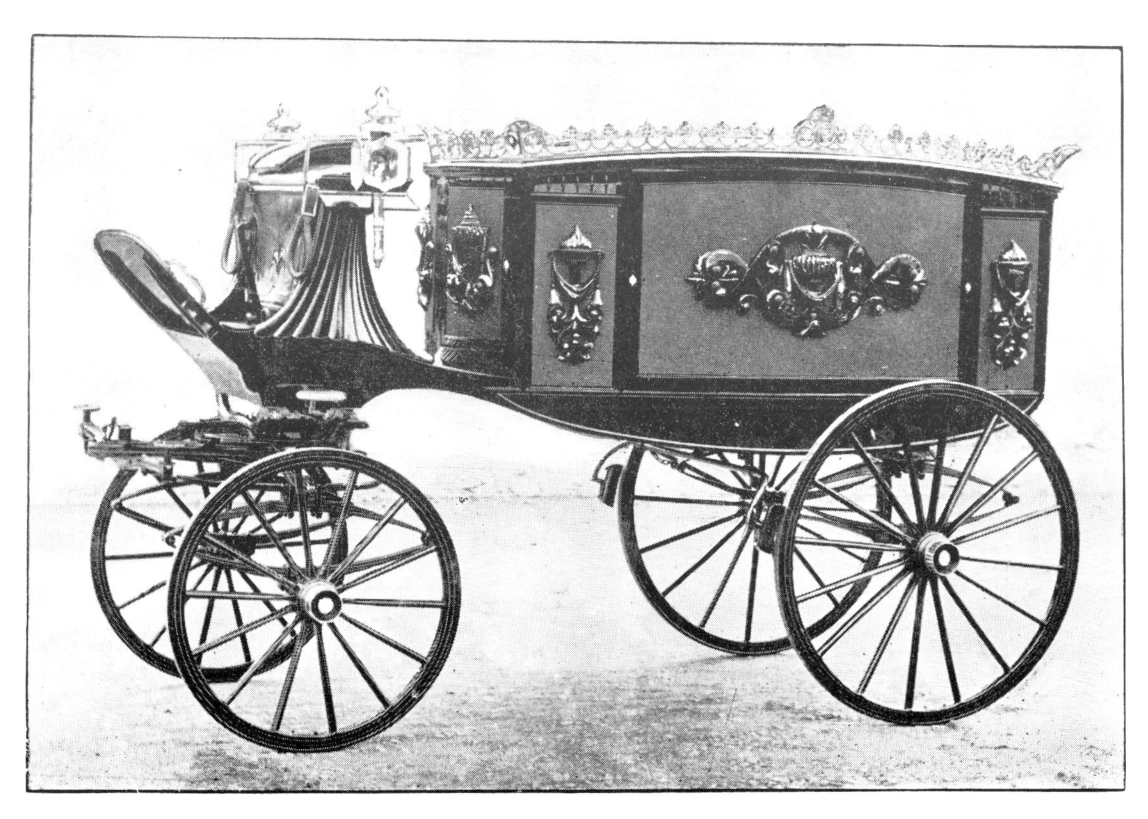

No. 178. HEARSE.

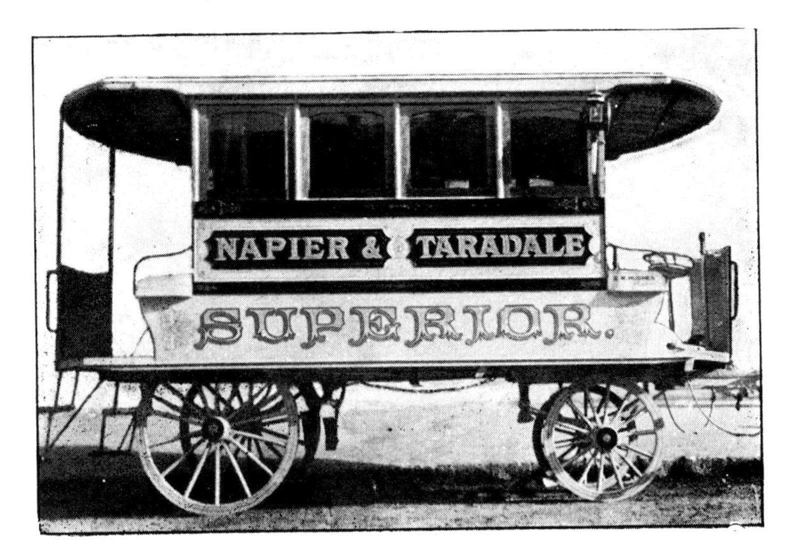

No. **179**. PALACE CAR.

No. **180**. EIGHTEEN PASSENGER DRAG.

No. 181. ROAD CART.

No. 182. FRUIT GROWER'S WAGGON.

No. **183.** COVERED DAIRYMAN'S CART.

No. **184.** DAIRYMAN'S CART.

No. 185. BAKER'S CART.

No. 186. DAIRYMAN'S DELIVERY WAGGON.

TABLE OF MEASUREMENTS

The proportions here given are those adopted by Australasian coachbuilders of repute. They are given as such, and are not necessarily suitable for all localities. Copies of those working drawings which are marked as having appeared in the *Coachbuilder* may be obtained from the publishers at Sydney or Melbourne. Price 2/6 each.

No. 1.—VICTORIA. Body: widths — under rockers, 34in.; across door pillars, at seat, $41\frac{1}{2}$in.; top, 48in. Boot at front of bracket, 25in. Side quarters: length—at bottom side, 21in.; elbow-rail, 26in.; depth, $11\frac{1}{2}$in. Driving seat, 13in. x 35in. Length over body bottom, including bracket, 94in. Springs: hind, length over all, 48in.; 5 plates, $1\frac{1}{2}$in. x $\frac{1}{4}$in.; main plate, $\frac{5}{16}$in.; front, 37in.; 4 plates, $1\frac{1}{2}$in. x $\frac{1}{4}$in.; main plate, $\frac{5}{16}$in.; compass inside, 8in. Axles: Collinge, $1\frac{1}{4}$in.; over collars, 46in. and 42in. Wheels: dia., 42in. and 34in; spokes, $1\frac{3}{8}$in.; hubs, $5\frac{1}{2}$in. x $7\frac{1}{2}$in.; rims, $1\frac{1}{2}$in; tyres, $1\frac{3}{8}$in. x $\frac{7}{16}$in.

No. 2.—LANDAU. Body: widths at elbow-line, front, 40in.; back, 39in.; across door pillars at seat, 50in.; top, 55in.; Boot, front, 30in.; length on elbow-line, front, $21\frac{1}{2}$in., back, $22\frac{1}{2}$in.; doorway, $21\frac{1}{2}$in. Boot: depth, 21in. Driving seat, $13\frac{1}{2}$in. x 37in. Length of body over all, 116in. Springs: hind, length over all, 48in.; 6 plates, $1\frac{3}{4}$in. x $\frac{1}{4}$in.; main plate, $\frac{5}{16}$in.; front, 37in. centres; 4 plates, $1\frac{1}{2}$in. x $\frac{1}{4}$in.; main plate. $\frac{5}{16}$in.; cross, 40in.; 6 plates, $1\frac{3}{4}$in. Axles: Collinge, $1\frac{1}{2}$in.; over collars, 48in Wheels: dia., 44in. and 37in; spokes, $1\frac{1}{2}$in.; hubs, $5\frac{3}{4}$in. and 6in. x 8in.; rims, $1\frac{3}{8}$in.; tyres, $1\frac{1}{2}$in. x $\frac{7}{16}$in.

No. 3.—SINGLE BROUGHAM. Body: widths—across bottom, 42in.; door pillars, 48in.; back quarter, top, 41in.; length on elbow-line, 44in.; doorway, $20\frac{1}{2}$in.; side panels, depth over mouldings, $13\frac{3}{4}$in.; length of body over bottom, including bracket, 88in. Boot, depth, 18in; seat, $12\frac{1}{2}$in. x 32in. Springs: hind, 38in. centres; 4 plates, $1\frac{3}{8}$in. x $\frac{1}{4}$in.; compass inside, 7in.; front, 36in.; 4 plates, $1\frac{1}{2}$in. x $\frac{1}{4}$in.; compass inside, 8in. Axles: Collinge, $1\frac{1}{4}$in.; over collars, 46in. Wheels, 44in. and 37in.; spokes, $1\frac{3}{8}$in.; hubs, $5\frac{1}{2}$in. x $7\frac{1}{2}$in.; rims, $1\frac{3}{8}$in.; tyres, $1\frac{1}{2}$in. x $\frac{7}{16}$in.
Working drawing in *Coachbuilder*, Vol. 5, No. 7.

No. 4.—BROUGHAM. Body: widths—across bottom, 44in.; top, 51in.; front of boot, 31in.; back of body, $42\frac{1}{2}$in.; box-seat, 36in. x 14in.; length on elbow-line 46in.; depth of panels over mouldings, 15in.; doorway, 22in.; length over all, including bracket, 96in. Springs: hind, centres, 39in.; 5 plates, $1\frac{3}{8}$in x $\frac{1}{4}$in.; main plate, $\frac{5}{16}$in.; front, 39in.; 4 plates, $1\frac{1}{2}$in. x $\frac{1}{4}$in; main plate, $\frac{5}{16}$in. Axles: Collinge, $1\frac{3}{8}$in.; over collars, 46in. Wheels: 47in. and 36in.; spokes, $1\frac{1}{2}$in.; hubs, $5\frac{1}{4}$in. and $5\frac{1}{2}$in. x $7\frac{1}{2}$in.; rims, $1\frac{1}{2}$in.; tyres, $1\frac{3}{8}$in. x $\frac{3}{8}$in.

No. 5.—WAGGONETTE (private). Body: widths—across bottom, 30in.; front of bracket, $29\frac{1}{2}$in.; under front seat, 35in.; length over bottom, including bracket, 78in.; cut-under, length, 24in. Seats: front seat, bottom, 15in. x 38in.; across elbows, front, 43in.; rear seats, 13in. x 38in. on bottom; between seats, $18\frac{1}{2}$in. Springs: hind, 36in.; 5 plates, $1\frac{1}{2}$in. x $\frac{1}{4}$in.; compass inside, $7\frac{1}{2}$in.; front, 34in.; 4 plates, $1\frac{1}{2}$in. x $\frac{1}{4}$in.; compass inside, 7in. Axles: $1\frac{1}{4}$in. and $1\frac{3}{8}$in.; over collars, 52in. and 46in. Wheels, 50in. and 42in. dia.; hubs, $5\frac{3}{4}$in. x 8in.; spokes, $1\frac{1}{2}$in.; rims, $1\frac{3}{8}$in. x $1\frac{1}{2}$in.; tyres, $1\frac{3}{8}$in. x $\frac{7}{16}$in.
Working drawing in the *Coachbuilder*, Vol. 6, No. 7.

No. 6.—WAGGONETTE (Stanhope). Body: widths—across bottom, 31in.; top, 34in.; depth of side panels, $9\frac{3}{4}$in.; length of bottom, including bracket, 77in.; across cut-under, 22in. Front seat, frame, over, 41in. x 17in.; top of pillars, 48in.: depth of rails, 11in.; rear seats, frame, 36in. x 14in.; width between, 19in. Springs: hind, 38in.; 6 plates, $1\frac{1}{2}$in. x $\frac{1}{4}$in.; compass inside, $8\frac{1}{2}$in.; front, 34in.; 4 plates, $1\frac{1}{2}$in. x $\frac{1}{4}$in.; compass inside, 8in. Axles: Collinge, $1\frac{1}{2}$in. and $1\frac{3}{8}$in.; over collars, 52in. and 46in. Wheels, 50in. and 40in.; spokes, $1\frac{1}{2}$in.; hubs, $5\frac{3}{4}$in. and $5\frac{1}{2}$in. x $7\frac{1}{2}$in.; rims, $1\frac{1}{2}$in.; tyres, $1\frac{3}{8}$in. x $\frac{7}{16}$in.
Working drawing in the *Coachbuilder*, Vol. 8, No. 2.

No. 7.—PONY PHAETON (open). Widths across rockers, 30in.; seats, bottom, 38in.; top, 44in.; depth of seats, front, 13in.; rear, 18in.; between seats, 21in.; depth of panels, hind, 11in. Springs: hind, 36in.; 4 plates $1\frac{1}{4}$in. x $\frac{1}{4}$in.; front, 32in.; 4 plates, $1\frac{1}{2}$in. x $\frac{1}{4}$in. Axles, $1\frac{1}{8}$in.; over collars, 44in. Wheels, 42in. and 30in.; spokes, $1\frac{1}{4}$in.; tyres, $1\frac{1}{8}$in. x $\frac{5}{16}$ in.

No. 8.—PONY PHAETON (small size). Width across rockers 30in.; rear seat on frame, 17in. x 36in.; across top 42in.; front seat, 14in. x $35\frac{1}{2}$in.; between seats, 22in.; depth of seat panels over mouldings, $9\frac{1}{2}$in. Springs, hind, 36in.; 4 plates, $1\frac{1}{4}$in. x $\frac{1}{4}$in.; front, 34in.; 3 plates, $1\frac{1}{4}$in. x $\frac{1}{4}$in. Axles, $1\frac{1}{8}$in.; length, 40in. Wheels, 38in. and 28in.; spokes, $1\frac{1}{4}$in.; tyres, $1\frac{1}{8}$in. x $\frac{5}{16}$in.

No. 9.—PONY PHAETON (hooded). Width across rockers, 30in.; depth of rockers, 12in.; rear seat, outsides bottom frame, 38in. x 17in.; depth of panels, 10in.; width across top, 43in.; length on elbow rails, 21in.; front seat, frame, $37\frac{1}{2}$in. x 14in.; between seats, 21in. Springs: hind, 36in.; 4 plates, $1\frac{1}{4}$in. x $\frac{1}{4}$in.; compass inside, 8in.; front, 32in.; 4 plates, $1\frac{1}{2}$in. x $\frac{1}{4}$in.; compass inside, 8in. Axles, $1\frac{1}{8}$in.; length, 42in. Wheels, 42in. and 30in.; spokes, $1\frac{1}{4}$in.; hubs, 5in. x 7in.; rims, $1\frac{1}{4}$in.; tyres, $1\frac{1}{8}$in. x $\frac{5}{16}$in.
Working drawing appeared in *Coachbuilder*, Vol. 9, No. 9.

No. 10.—PONY PHAETON (hooded). Measurements same as No. 7.

No. 11.—PONY PHAETON (open). Width over rockers, 31in.; depth rockers, 10in. Seats: rear, 36in. x 17in.; across top, 42in.; elbow rail, 21in.; front seat, 36in. x 14in.; between seats, 22in. Springs and axles, same measurements as No. 9. Wheels, 30in. and 42in.; spokes, $1\frac{1}{8}$in.

No. 12.—PONY PHAETON (English carriage). Body, same measurements as No. 11. Springs: hind, 36in.; 4 plates, $1\frac{1}{4}$in. x $\frac{1}{4}$in.; compass inside, 8in.; front, 34in.; 3 plates, $1\frac{1}{4}$in. x $\frac{1}{4}$in.; compass inside, 8in. Axles, $1\frac{1}{8}$in.; length, 44in. and 40in. Wheels, 32in. and 43in.; spokes, $1\frac{1}{4}$in.
Working drawing appeared in the *Coachbuilder*, Vol. 9, No. 9.

No. 13.—VILLAGE PHAETON Body: widths—bottom, 31in.; top, 38in.; length on bottom, 57½in.; depth of sides over rails, 19½in.; depths of seats: front, 13in.; hind, 15in. Springs: hind, 36in ; 4 plates 1½in. x ¼in.; compass inside, 8in.; front, 32in.; 3 plates, 1¼in. and ¼in.; compass inside, 8½in. Axles, 1⅛in.; length, 44in. and 38in. Wheels, 42in. and 30in.; spokes 1¼in. and 1⅜in.

Working drawing appeared in the *Coachbuilder*, Vol. 8, No. 4.

No. 14—PONY PHAETON (hooded). Width—across rockers, 30in.; seats, hind, width, bottom, 36in.; top, 42in.; depth, hind seat, 17½in.; front, 15in ; depth of panel, 10in. Springs and axles, same as No. 11. Wheels, 46in. and 34in.

No. 15.—PHAETON (open, drop front). Width across bottom, 28in.; seat, width, bottom, 36in. x 17½in.; top, 40in.; depth of panels over all, 10in.; depth of rockers, 11in.; length over bottom, including bracket, 41½in. Springs, hind, 34in.; 3 plates, 1¼in. x ¼in.; front, 34in.; 4 plates, 1¼in. x ¼in ; compass, 8in. and 7½in. inside. Length over undercarriage, 52in.; Axles, 1in.; length, 44in. Wheels, 44in. and 36in.; spokes, 1⅛in.; hubs, 4½in. x 6½in.

No. 16.—PHAETON. (Hooded drop front.) Same measurements as No. 15.

No. 17.—OPEN PHAETON. Width across front of seat, bottom, 36in.; top, 40in.; across bottom, 32in.; depth of seat-panels, 12in.; length on elbow, 22in.; length over body, bottom, 49 in. Undercarriage, springs and axles same as No. 15, excepting diameter of wheels, which are 42in. and 46in.

No. 18.—BUGGY. Extension top, two seat. Length, bottom, 64in.; depth of sides, 12in.; seat riser, 3¼in. Widths, underneath, 30in.; top, 33in. Seats: frame, 36in. x 18in.; leg room, 24½in.; between seats, 14in. Springs: front, 36in.; 4 plates, 1½in. x ¼in.; main plate, $\frac{5}{16}$in.; compass inside, 7in.; hind, 36in.; 5 plates, 1½in. x ¼in.; main plate, $\frac{5}{16}$in.; compass, 7½in. Wheels: 48in. and 42in; spokes, 1¼in. Axles: 1⅛in.; length, 50in.

No. 19.—DOG CART. Full size, four wheel. Body: widths—across bottom, 33in.; top, 37in.; depth of sides, 15¾in.; length over all, including bracket, 77½in.; bracket, 14in.; set-up; at point, 8¼in. Seats: front, 39in. x 15½in.; hind, 39in. x 14½in.; leg room, 25in. Springs: hind, 37in.; 4 plates, 1½in. x ¼in, main plate, $\frac{5}{16}$in.; front, 35in.; 4 plates, 1⅜in. x ¼in.; main plate, $\frac{5}{16}$in.; compass inside, both, 7½in. Axles: 1¼in.; collinge, length over collars, 48in. Wheels: 50in. and 36in.; spokes, 1½in.; hubs, 7¾in. x 5¾in.; rims, 1⅝in.; tyres, 1¼in. x ⅜in.

Working drawing appeared in the *Coachbuilder*, Vol. 9, No. 6.

No. 20.—STATION TRAP. Body: widths—bottom, 35in.; top, 39in.; depth of sides, 15in.; length of body, including bracket, 65in.; bracket, 14½in.; set up, 7½in. Seats: width, bottom, 42in.; top, 44½in.; top panels: length, 34in.; depth, 7½in. Undercarriage: length, 58in.; axle beds, 1½in. wide; side reaches 1¼in. square; middle reach, 1¼in. x 1⅜in.; wheel plate dia., 18in. Springs: side, 52in.; 6 plates, 1⅜in. x ¼in.; main p'ate, $\frac{5}{16}$in.; compass inside to centres, 8in. Axles, 1¼in.; length, 52in. Wheels, 50in. and 44in.; spokes, 1⅜in.; hubs, 5¼in. x 7¼in.; rims, 1⅜in.; tyres, 1¼in.

Working drawing appeared in the *Coachbuilder*, Vol. 11, No. 5.

No. 21.—DOG CART (curved panel, 4 wheel). Body: widths—bottom, 31in.; top, 35in.; over curved panels, 47in.; depth of side panels, 12in.; curved ditto., 11in. deep, and 38in. long on bottom; length of body, including bracket, 70in. Springs: hind, 38in.; 4 plates, 1¼in. x ¼in.; compass inside, 7in.; front, 34in.; 3 plates, 1¼in. x ¼in.; compass inside, 6½in. Axles, 1¼in.; length, 44in. Wheels, 46in. and 36in.; 1¼in. spokes.

Working drawing appeared in the *Coachbuilder*, Vol. 8, No. 8.

No. 22.—DOG CART (cut under). Body: widths—bottom, 31in.; top, 35in.; over top panels, 38in.; side panel, depth, 12in.; top panel, 5¾in. x 34in.; length over all, including bracket, 64in. Springs: hind, 36in.; 3 plates, 1½in. x ¼in.; compass inside, 8in.; front, 34in.; 4 plates, 1½in. x ¼in.; compass, 8in. Axles, 1⅛in.; length, 50in. Wheels, 50in. and 42in.; spokes, 1⅜in.

No. 23.—NUNDAH TRAP. Body: widths—bottom, 31in.; top, 34in ; depth of sides, 14½in ; length, bottom, including bracket, 56½in. Springs: hind, 34in.; 3 plates, 1¼in. x ¼in.; compass inside, 6½in.; front, 34in.; 4 plates, 1½in. x ¼in.; compass, 6½in. Axles, 1⅛in.; h.p., length, 48in. Wheels, 46in. and 40in.; spokes, 1¼in.; hubs, 5in. x 7in.; rims, 1⅜in.; tyres, 1⅛in. x ⅜in.

Working drawing appeared in the *Coachbuilder*, Vol. 7, No. 3.

No. 24.—DOG CART (slatted, on side springs). Body: widths—bottom, 32in; top, 36in.; depth of sides, 15in.; length, top of top batten, 35in.; length ; body over all, including bracket, 66in. Undercarriage, length, 58in. Side springs, 52in.; 6 plates, 1⅜in. x ¼in.; main plate, $\frac{5}{16}$in.; compass inside, 8in.; cross spring, 34in.; 3 plates, 1½in. x ¼in. Axles, 1¼in.; h.p. length, 52in. Wheels, 50in. and 42in.; spokes, 1⅜in.; hubs, 5½in. x 7½in.; rims, 1½in.; tyres, 1¼in. x ½in.

No. 25.—TRAP (four passenger). Body: widths—bottom, 30in.; top, 33½in.: across top rails, 36in.; length, bottom, including bracket, 57in.; depth of side panels, 12in. Undercarriage and wheels same as No. 23.

No. 26.—CANOPY TOP TRAP. Body: widths—bottom, 32in.; top, 35in ; over top panels, 38in.; length over all, 67in. Springs: hind, duplex, 46in.; 4 plates, 1¼in. x ¼in.; front, single, 44in.; 4 plates, 1½in. x ¼in. Axles, 1⅛in.; length, 50in. Wheels, 44in. and 48in.; sarven, spokes, 1¼in.

No. 27.—TRAY BUGGY (two seat). Body; bottom—width, 33in.; length, without bracket, 57in.; depth of side under front seat, 19in. Seats, front, bottom frame, 38in. x 15in.; back, 36in. x 14in. Undercarriage: length, 58in. Springs: side, 52in.; 6 plates, 1⅜in. x ¼in.; main plate, $\frac{5}{16}$in. ; cross spring hung in lyre shackles, 44in.; 4 plates 1½in. x ¼in ; main plate. $\frac{5}{16}$ in. Axles, 1¼in.; h.p., length, 52in. Wheels 50in. and 44in.; spokes, 1½in.; hubs, 6in x 8in.; rims, 1⅜in.; tyres, 1¼in. x ½in.

No. 28.—DOUBLE ABBOTT BUGGY. Body: width—bottom, 35in.; length, bottom, 66in. Seats: over frame, front, 39in. x 14in.; back, 37in. x 15in. Wheels and undercarriage, same as No. 27. Springs: side, 52in.; 8 plates, 1⅜in. x $\frac{3}{16}$in.; main plate, ¼in.

Working drawing appeared in the *Coachbuilder*, Vol. 6, No. 2.

TABLE OF MEASUREMENTS.

No. 29.—TURNOUT SEAT BUGGY (hooded). Body; width—bottom, outsides, 32in.; top, 35in.; depth of sides, 12in.; length over all, 59in.; length, bottom, including bracket, 54in. Seats: front, 38in. x 16in.; back, 33in. x 12in. Undercarriage: length, 58in., 3 reaches; side springs, 52in.; 6 plates, 1½in. x ¼in.; compass inside, 8in.; cross spring, 34in.; 3 plates. 1¼in. x ¼in. Axles, 1¼in.; h.p., length, 52in. Wheels, 50in. and 44in.; spokes, 1½in.; hubs, 6in. x 8in.; rims, 1½in.; tyres, 1¼in. x ½in.

No. 30.—SLIDE SEAT BUGGY (hooded). Body: widths—bottom, outsides, 32in.; top, 36in.; depth of panels, 7½in.; depth of riser, 3in.; length of body over all, 58in.; on bottom, including bracket, 52in. Seat over frame, 40in. x 18in.; jump seat, 32in. x 12in. Undercarriage, 58in Side springs, 52in.; 6 plates, 1¾in. x ¼in.; opening inside to centres, 7½in.; cross springs in lyre shackles, 44in.; 4 plates, 1½in. x ¼in. Axles 1⅛in.; h.p. length, 52in. Wheels, 50in. and 44in.; spokes, 1⅜in.; hubs, 5½in. x 7½in.; rims, 1½in.; tyres, 1¼in. x $\frac{7}{16}$in.

Working drawing appeared in the *Coachbuilder*, Vol. 3, No. 3.

No. 31.—GODDARD BUGGY. Body: widths—across rockers, 28in.; under body at back, 34½in.; at point of bracket, 30in.; over body, 37in. Seat over bottom frame, 37in. x 20½in.; width over elbows, 45in.; depth of seat panels, 6in.; length of body over all, 52in.; depth over rockers, 13in. Undercarriage, length 59in. Springs; front, 36in.; 4 plates, 1½in. x ¼in.; compass inside, 8in.; back, 37in.; 4 plates, 1½in. x ¼in.; compass, 8½in. Wheels, 48in. and 44in.; spokes, 1¼in.; hubs, 4¼in. x 6½in.; rims, 1¼in. Axles, 1in.; h.p., length, 52in.

Working drawing appeared in the *Coachbuilder*, Vol. 6, No. 1.

No. 32.—ABBOTT BUGGY (hooded). Tray: bottom frame over mouldings, 56in. x 32in. Seat: bottom frame inside mouldings, 36in. x 16in.; across elbows, 42in.; leg room, 27in. Undercarriage, 58in.; 3 reaches; side springs, 52in.; 5 plates, 1⅜in. x ¼in.; compass inside to centres, 7in. Axles, 1⅛in.; h.p., length, 52in. Wheels, 50in. and 44in.; spokes, 1⅜in.; hubs, 5¼in. x 7¼in.; rims, 1½in.; tyres, 1⅛in. x $\frac{7}{16}$in.

Working drawing appeared in the *Coachbuilder*, Vol. 7, No. 6.

No 33.—PIANO BOX BUGGY (on side bars). Body: widths—bottom, 24in.; top, 25in.; length bottom, 52in.; depth of panels, 7¾in.; of risers, 3½in. Seat, bottom frame, 32in. x 16in. Springs: back and front, inside side bars, 28in.; 3 plates, 1¼in. x ¼in. Body springs 29in.; 3 plates, 1¼in. x ¼in. Axles, 1in.; length, 48in. Wheels, 48in. and 46in.; spokes, 1⅛in.; rims, 1¼in.

No. 34.—PIANO BOX BUGGY (hooded, on Dexter springs). Body: width—on bottom, 27in.; length, 52in.; depth of side panels, 8in.; do. of seat risers, 3¾in.; seat, 16½in. x 36in. Springs, 55in; 4 plates, 1¼in. x ¼in. Axles, 1⅛in.; h.p., length, 50in. Wheels, 46in. and 42in.; spokes, 1⅛in.

No. 35.—PIANO BOX BUGGY (on elliptic springs). Body: width—on bottom, 26in.; length, 51in.; depth of panels, 8in.; risers, 3½in. Seat, bottom frame, 34in. x 16in. Springs, elliptic, 36in.; 3 plates, 1½in. x ¼in.; compass inside; front, 6in.; back, 6½in. Wheels and axles, same as No. 33.

No. 36.—PIANO BOX BUGGY (hooded). Body: width—over bottom, 28in.; length, 54in.; depth of side panels, 8½in; depth of riser. 3½in; seat bottom over frame, 16½in. x 34in. Springs, 50in.; 5 plates, 1½in. x ¼in.; compass inside to centres, 5in. Axles and wheels, same as No. 33.

No. 37.—COAL BOX BUGGY (hooded). Body: bottom width across rockers, 25in.; over all, 30in.; over top panel, 35in. Seat over frame, 36in. x 17in.; length over all, 57in.; bottom, including bracket, 51in.; leg room, 27in Wheels and undercarriage, same as No. 32.

Working drawing appeared in the *Coachbuilder*, Vol. 8, No. 8.

No. 38.—SINGLE BUGGY (hooded). Body: width—bottom, 30in.; length, 50in.; seat over frame, 34in. x 17in. Wheels and undercarriage, same as No. 35.

No. 39.—CONCORD BUGGY. Body: tray over bottom, 52in. x 26in.; seat over frame, 34in. x 16½in. Undercarriage, 56in. Springs, 50in.; 5 plates, 1½in. x ¼in.; compass inside to centres, 5in. Axles, 1in.; h.p., length, 48in. Wheels, 44in. and 40in.; spokes, 1⅛in.

No. 40.—CORNING BUGGY (on Dexter Queen Springs.) Body: width—bottom, 26in.; length, 49in.; depth of sides, 12in. Seat over frame, 32in. x 15in. Undercarriage: length, 56in. Springs, 54in.; 4 plates, 1¼in. x ¼in. Wheels, 46in. and 44in.; spokes, 1⅛in. Axles, 1in.; length, 48in.

No. 41.—WHITECHAPEL SPEEDING BUGGY. Body: width—bottom, 25in.; top, 28in.; length, bottom, 53in.; depth of sides, 13in. Seat: over bottom frame, 32in. x 16in.; across top, 37in. Springs: 43in., 1¼in. x ¼in.; front, 3 plates; hind, 4 plates. Axles, h.p., 1⅛in.; length, 50in. Wheels, 48in. and 44in.; sarven, spokes, 1¼in.

No. 42.—COAL BOX BUGGY. Body: width—bottom, 30in.; top, 34in.; depth of side panels, 11in.; length, including bracket, 50in. Undercarriage and wheel, same as No. 37.

No. 43.—MARNI BUGGY. Body: width—bottom, 23in.; top, 24in.; length bottom. 54in.; depth of side panels, 8in.; seat risers. 3¾in. Seat outside, bottom frame, 32in. x 16in. Undercarriage and wheels, same as No. 41.

No. 44.—WHITECHAPEL BUGGY (on elliptic springs). Body: width—bottom, 25in.; top, 27in.; length, bottom, 48in.; depth of sides, including riser, 11in. Seat over bottom frame 32in. x 16in. Springs: both 36in.; 4 plates, 1⅜in. x ¼in.; compass inside. 6in. and 6½in. Wheels, 44in. and 40in.; spokes, 1⅛in.; hubs, 6½in. x 4in.; rims, 1¼in.

No. 45.—COAL BOX BUGGY (open). Measurements, same as No. 37.

No. 46.—ABBOTT BUGGY (open). Measurements, same as No. 32.

No. 47.—SINGLE BUGGY. Body: width—bottom, 30in.; top, 33in.; length, bottom, 51in.; over all, 58½in.; depth of side panels, including riser, 11in. Seat on bottom frame, 36in. x 16in.; across top, 40in. Undercarriage: 3 reaches; length, 57in. Springs, 50in.; 5 plates, 1½in. x ¼in.; compass inside to centres, 7in. Axles: h.p., 1⅛in.; length, 50in. Wheels, 50in. and 44in.; sarven; 1⅜in. spokes.

Working drawing appeared in the *Coachbuilder*, Vol. 9, No. 9.

No. 48.—Excelsior Buggy (single). Body: width—bottom, 31in.; top, 34in.; length, bottom, 52in. Seat: bottom frame, 36in. x 15½in.; across elbow rails, 44in. Undercarriage, and wheels same as No. 47.

No. 49.—Single Seat Buggy. Body: width—bottom, 31in.; top, 34in.; length, bottom, including bracket, 52in. Seat: bottom frame, 38in. x 15¼in. Undercarriage and wheels, same as No. 47.

No. 50.—Open Piano Box Buggy (with spindle seat) Body: width—bottom, 24in; top, 24¾in.; length, bottom, 64in.; depth of sides, 8in.; seat risers, 3¼in Seat: bottom frame, 32in. x 16½in.; depth over side rails, 8in. Springs; 3 plates, 1¼in. x ¼in; between side bars, 29in. Axles: 1in., h.p., length, 50in. Wheels, 48in. and 44in.; spokes, 1⅛in.

No. 51.—Single Whitechapel Buggy. Body: width—bottom, 30in.; top, 33in.; depth of sides, 12½in.; length, bottom, 50in. Seat: bottom frame, 36in. x 16in.; across top, 40in. Undercarriage and wheels, same as No. 32.

No. 52.—Piano Box Buggy (open). Body: width—bottom, 25in.; top, 26½in.; depth of sides, 7¾in.; seat riser, 3½in.; length, bottom, 60in. Springs, 29in.; 4 plates, 1¼in. x ¼in. Axles, 1in.; length, 48in. Wheels, 48in. and 45in.; spokes, 1⅛in.

No. 53.—Speeding Buggy. Measurements same as No. 41, excepting length of body on bottom, 57in.

Working drawing appeared in the *Coachbuilder*, Vol. 4, No. 6.

No. 54.—Corning Buggy (on side springs). Body: width—bottom, 24½in.; top, 26in.; length, bottom, 49in. Seat over bottom frame, 29in. x 17in.; over top, 36in. Undercarriage, 57in. Springs, 50in.; 5 plates, 1½in. x ¼in.; compass inside, 4½in. Axles, 1⅛in., h.p., length, 50in. Wheels, 45in. and 48in.; spokes, 1⅛in.

No. 55.—Speeding Buggy (with boot). Measurements same as No. 41, excepting length of body on bottom, 58in.

No. 56.—Convertible Seat Buggy (light). Body: width—bottom, 32in.; top, 35in.; depth of sides, 12½in; length, bottom, 50in. Seat: bottom frame, 38in. x 16½in.; across top, 42in. Undercarriage, 4ft. 11in. Springs, 52in.; 6 plates, 1¾in. x ¼in. Axles, 1¼in. Wheels, 50in. and 44in.; spokes, 1⅛in.

Nos. 57 and 58.—Franklin Buggy. Body: width—bottom, 32in.; top, 35in.; length over all, 68in.; depth of sides under seat, 13½in.; seat riser, 5in. Seats: front, 17in. x 39in.; back, 13in. x 32in. Undergear: three reaches, side, 1¼in. square; middle, 1¼in. x 1⅜in.; beds, 1½in. wide; headblock, 2½in. x 1¾in. Springs: side, 52in.; 6 plates, 1¾in. x ¼in.; compass inside to centres, 8in.; back, 44in.; 4 plates, 1½in. x ¼in. Wheels, 50in. and 44in.; spokes, 1⅛in.; hubs, 5½in. x 7½in.; rims, 1⅜in.; tyres 1¼in. x $\frac{7}{16}$in. Axles, 1¼in. h.p., length, 52in.

Nos. 59 and 60.—Convertible Buggy Waggon. Body: width—bottom, 32in.; top, 35in.; length over bottom, including bracket, 62in.; depth of sides, 13½in. Seat, 16in. x 38in. Undergear and wheels, same as No. 57.

Nos. 61 and 62.—Convertible-Seat Buggy Waggon. Body: width—bottom, 30in.; top, 34in.; length, bottom, including bracket, 65in.; depth of sides, 8¾in.; seat riser, 5½in. Seat on bottom frame, 38in. x 17in. Undergear, same as No. 57.

Working drawing in *Coachbuilder*, Vol. 5, No. 6.

No. 63.—Shellback Turnout-Seat Buggy. Body: width—bottom, 32in.; top, 35in.; length, bottom, including bracket, 56in.; depth of side panels, 12in. Seats: front, 38in. x 16in.; back, 13in. x 31in. Undergear, 3 reaches. Springs: side, 52in.; 5 plates, 1¾in. x¼in.; compass, 8in.; back, 36in.; 3 plates, 1½in. ¼in. Axles, 1¼in.; length, 52in. Wheels, 50in. and 44in.; spokes, 1⅛in.

No. 64.—Turnover Seat Buggy. Body: depth of side panels under front seat, 14in.; back, 8½in. All other measurements same as No. 63, excepting back spring, elliptic, 34in.; 3 plates, 1½in. x ¼in.

No. 65.—Turnout Seat Buggy. Same as No. 63.

No. 66.—Turnout Seat Buggy. Width—bottom, 32in.; top, 35in.; length, bottom, including bracket, 57in.; depth of sides, 14in. Seats: front, 39in. x 16in.; back, 12½in. Undergear same as No. 63, excepting back spring, 34in.; 3 plates, 1½in. x ¼in.

No. 67.—Turnout Seat Buggy. Body: width over bottom, 32in.; top, 35in.; length, bottom, including bracket, 54in. Seats: front, 16in. x 36in.; back, 12½in. Undergear same as No. 63, excepting back spring, single plate, elliptic, 34in.

No. 68.—Jump-Seat Buggy. Back seat on jump-irons Body: width, bottom, 32in.; top, 35in.; depth of sides, 14in; length, including bracket, 56in. Seats: front, 39in. x 16in.; back, 13in. Other measurements same as No. 57.

No. 69.—Turnout Seat Buggy. Body: depth of sides, 14in.; length on bottom, including bracket, 54in.; over all, 60in. Other measurements same as No. 64.

No. 70.—Jump-Seat Buggy. Body: widths—bottom, 32in.; top, 35in.; length, bottom, including bracket, 54in. Back seat, on bottom, 35in. x 17in.; jump seat, 29in. x 12in. Jump irons to centres, front, 15in.; back, 11in.; depth of body side panels, 9in.; seat risers, 2½in. Undercarriage same as No. 57.

Working drawing appeared in the *Coachbuilder*, Vol. 7, No. 9.

No. 71.—Angus Buggy. Tray: length of bottom, 56in; width, 32in. Seat: bottom frame, 34in. x 15½in.; across elbows, 40in. Undercarriage, 3 reaches; length, 57in. Springs: side, 50in.; 5 plates, 1½in. x ¼in; compass inside, 7in.; cross spring, 44in.; 3 plates, 1½in. x ¼in. Axles: 1⅛in.; h.p., length, 50in. Wheels: 50in. and 44in., sarven; spokes, 1⅛in.

No. 72.—Farmer's Waggonette. Body: width—bottom, 34½in.; top, 39in.; length, bottom, including bracket, 70in.; bracket, 12in.; set-up, 5½in.; leg room, 26in. Front seat, on bottom, 44in. x 16in.; back seats, 33in. x 14in.; between seats, 18in. Undercarriage: 59in.; side springs, 52in. x 1¾in., 6 plates; back spring, 34in. x 1½in., 4 plates; thickness of plates, main, $\frac{5}{16}$in.; others, ¼in. Axles: 1¼in.; length, 52in. Wheels: 50in. and 44in.; spokes, 1½in.; hubs, 6in. x 8in.; rims, 1⅝in. Body to ground, 40in.

Working drawing in *Coachbuilder*, Vol. 6, No. 4.

No. 73.—Two Seat Excelsior Buggy. Body: width—bottom, 33in.: top, 36in.; length, bottom, 62in.; leg room, 27in.; depth of sides, 14in. Seats: front, 38in. x 16in.; back, 38in. x 14½in. Undergear same as No. 64.

TABLE OF MEASUREMENTS.

No. 74.—Farmer's Buggy Waggon. Body: width—bottom, 32in.; top, 35in.; length, bottom, including bracket, 61in.; bracket, 13in.; set-up, 5in.; leg room, 26in.; depth of sides, 13in. Seats: front, 41in. x $16\frac{1}{2}$in.; back, depth, 14in. Wheels: 50in. and 44in.; $1\frac{1}{2}$in. spokes. Undercarriage same as No. 57.

Working drawing appeared in *Coachbuilder*, Vol. 8, o. 2.

No. 75.—Two-Seat Open-Sided Express. Width—bottom, 31in.; top, 34in.; length, bottom, including bracket, $66\frac{1}{2}$in.; depth of sides under front seat, 13in.; depth of seat-risers, front, $4\frac{1}{2}$in.; back, $3\frac{1}{2}$in. Seats: 36in. x $15\frac{1}{2}$in. Springs: side, 32in., $1\frac{1}{2}$in x $\frac{1}{4}$in., 3 plates; compass inside, 7in.; front, 32in., $1\frac{1}{2}$in. x $\frac{1}{4}$in., 4 plates; compass, 7in. Axles: $1\frac{1}{4}$in.; length, 50in. Wheels: 50in. and 44in.; spokes, $1\frac{3}{8}$in.

No. 76.—Farmer's Two-Seat Waggon. Body: width—bottom, 34in.; top, 36in.; length, bottom, 67in.; leg room, 26in.; depth of sides under front seat, 17in.; under back seat, 14in. Seats: on bottom, 40in. x 16in. Undercarriage; 2 reaches, length, 59in. Springs: side, 34in., $1\frac{1}{2}$in. x $\frac{1}{4}$in., 4 plates: compass inside, 8in.; front, 34in., $1\frac{3}{4}$in. x $\frac{1}{4}$in., 5 plates; compass, 8in. Axles: $1\frac{1}{4}$in. Wheels: 50in. and 44in.; spokes, $1\frac{1}{2}$in.

No. 77.—Bracket-Front Buggy Waggon. Body: width—bottom, 32in.; top, 35in.; length, bottom, including bracket, 64in.; depth of sides, $13\frac{1}{2}$in. Seats: front, 38in. x 16in.; back, 38in. x 15in. Undercarriage same as No. 78, omitting front spring.

No. 78.—Squatter's Express. Body: width—over frame, 36in.; length, 72in.; depth of sides, 8in.; depth of seat risers, front, 10in.; hind, 6in. Seats: 40in. x 16in. Undercarriage: 3 reaches, 58in. Springs: side, 52in., 6 plates, $1\frac{3}{4}$in. x $\frac{1}{4}$in.; compass inside to centres, 8in.; back, 44in.; 4 plates, $1\frac{1}{2}$in x $\frac{1}{4}$in.; front, 36in.; 3 plates, $1\frac{1}{2}$in. x $\frac{1}{4}$in. Axles: $1\frac{1}{4}$in.; h.p., length, 52in. Wheels: 50in. and 44in.; spokes, $1\frac{1}{2}$in.; hubs, 6in. x 8in.; rims, $1\frac{5}{8}$in.

No. 79.—Adelaide Express. Body: width—bottom, 34in.; top, 36in.; length over all, 70in. Bracket: set-up, $6\frac{1}{2}$in.; length, 13in. Seats, spindle, 38in. x 16in. Undercarriage same as No. 63.

No. 80.—Farmer's Two-Seat Buggy-Waggon. Width—bottom, 32in.; top, 35in.; length, bottom, including bracket, 64in.; depth of sides over seat-risers, 13in.; seat-risers, depth, 5in. Seats: front, 38in. x 16in.; back, 38in. x 15in. Undercarriage same as No. 78, omitting front spring.

No. 81.—Farmer's Buggy Waggon. Body: width—bottom, 34in.; length, bottom, including bracket, 66in. Seats: 40in. x 16in. Undercarriage same as No. 72.

No. 82.—Canoe-Front Buggy-Waggon. Same as No. 86, excepting springs, cross spring, 34in.; 4 plates, $1\frac{1}{2}$in., x $\frac{1}{4}$in.; compass of side springs inside to centres, 6in.

Working drawing of body appeared in *Coachbuilder*, Vol. 9, No. 10.

No. 83.—Adelaide Express. Body: width—outside frame, 33in.; length, bottom, including bracket, 67in.; depth of side panels, 8in.; depth of seat risers, 6in. Seats, spindle, 38in. x 16in. Undercarriage same as No. 63.

Working drawing appeared in *Coachbuilder*, Vol. 5, No. 1.

No. 84.—Four-Passenger Trap. Body: width bottom, 32 in.; top 35in.; length over all, 71in. Seats: front, 38in. x 16in.; hind, 38in. x 15in. Undercarriage: 1 reach, length, 59in. Springs: sides, 34in.; 4 plates, $1\frac{1}{4}$in. x $\frac{1}{4}$in.; compass, inside, 8in.; front, 34in.; 4 plates, $1\frac{1}{2}$in. x $\frac{1}{4}$in.; compass, 8in. Axle, $1\frac{1}{8}$in.; length 52in. Wheels, 50in. and 44in.; spokes, $1\frac{3}{8}$in.

No. 85.—Run-About Buggy. Tray: 56in. x 26in. Seat: 36 in. x 16in.; seat to bottom, 14in. Undercarriage: length, 60in.; width across side bars, 30in. Springs, 28in.; 3 plates, $1\frac{1}{2}$in. x $\frac{1}{4}$in., or King Springs, $\frac{9}{16}$in. round. Axles: $1\frac{1}{8}$in.; length, 52in. Wheels: 48in. and 42in.; spokes, $1\frac{1}{4}$in.

No. 86.—Canoe-Front Buggy-Waggon. Body; width—bottom, 32in.; top, 35in.; depth of sides, 13in.; length bottom, including canoe front, 62in. Seats, 16in. and $15\frac{1}{2}$ in. x 38in. Undercarriage same as No. 57.

No. 87.—Buckboard. Tray, 26in. x 58in. Seat: bottom, 38in. x 16in. Axles: $1\frac{1}{8}$in., length, 52in. Wheels: 50in. and 44in.; sarven, $1\frac{3}{8}$in.

No. 88.—Queensland Buckboard. Tray: 9 battens, $2\frac{1}{4}$in. x $\frac{7}{8}$in.; length, 73in.; width across tray, 26in. Seat: 36in. x 16in.; box under seat, 26in. x $15\frac{1}{2}$in.; depth, 5in. Springs: 26in.; 2 plates, $1\frac{1}{4}$in. x $\frac{1}{4}$in.; compass inside, 7in. Axles: $1\frac{1}{4}$in.; h.p., length, 50in. Wheels: 50in. and 44in., $1\frac{3}{8}$in. sarven.

Working Drawing appeared in *Coachbuilder*, Vol. 7. No. 3.

No. 89.—Express-Waggon. Body: width—outside frame, 34in.; length, 72in.; depth of sides, 9in.; seat risers, $4\frac{1}{2}$in. Seat: 36in. x 16in. Springs: side, 34in.; 4 plates, $1\frac{1}{2}$in. x $\frac{1}{4}$in.; compass inside, 7in.; front, 34in.; 5 plates, $1\frac{1}{2}$in. x $\frac{1}{4}$in.; compass, 7in. Axles; $1\frac{1}{4}$in.; length, 52in. Wheels: 48in. and 42in.; spokes, $1\frac{3}{8}$in.

No. 90.—Open-Sided Express, with canopy top. Body: width—bottom, 32in.; length over bracket, 61in; depth of sides, $11\frac{1}{2}$in.; front seat riser, 3in. Seats: 38in. x 16in. Undercarriage: 3 reaches. Springs: side, 52in., 6 plates, $1\frac{3}{4}$in. x $1\frac{1}{4}$in.; compass inside to centre, 8in.; back, 34in.; 4 plates, $1\frac{1}{2}$in. and $\frac{1}{4}$in. Axles: $1\frac{1}{4}$in.; h. p., length, 52in. Wheels: 50in. and 44in.; spokes, $1\frac{3}{8}$in., sarven.

No. 91.—Express-Waggon. Body: length of bottom, 78in.; width, 39in.; depth of sides, $8\frac{1}{2}$in. Undercarriage, 2 reaches, length, 60in. Springs: front, 36in.; 6 plates, $1\frac{3}{4}$ in. x $\frac{1}{4}$in.; compass inside, 8in.; back, 36in.; 5 plates, $1\frac{1}{2}$ in. x $\frac{1}{4}$in.; compass inside, 8in. Axles: $1\frac{3}{8}$in.: length, 52 in. Wheels: 52in. and 44in.; spokes, $1\frac{1}{2}$in.; hubs, 6in. x 8in.; rims, $1\frac{5}{8}$in.; tyres, $1\frac{1}{4}$in. x $\frac{1}{2}$in.

No. 92.—Hampden Buggy. Body: width—bottom, 32in.; length, bottom, 65in.; depth of sides, 9in.; bevel back, 4in. Other measurements same as No. 90.

No. 93.—Express Waggon.—Same as No 91.

No. 94.—Station Waggon. Body: width—bottom, 40in.; length, bottom only, 72in.; bracket, 17in. Seat, 43in. x 16in. Undercarriage: 1 reach, length, 60in. Springs: side, 36in.; 5 plates, $1\frac{3}{4}$in. x $\frac{1}{4}$in.; main plate, $\frac{5}{16}$in.; compass inside, $8\frac{1}{2}$in.; front, 36in.; 6 plates, 2in. x $\frac{1}{4}$in.; main plate, $\frac{5}{16}$in.; compass, 8in. Axles, $1\frac{3}{8}$in.; length, 52in. Wheels: 52in. and 44in.; spokes, $1\frac{5}{8}$in.

No. 95.—PRINCE GEORGE CART. Body: width—bottom, 32in.; across top panels, 46in.; length, bottom, including bracket, 56in.; bracket, 17in.; set-up, 7in.; depth of side panels, 12in.; curved panels, depth, 13in. Springs: over all, 44in.; 4 plates, 1½in. x ¼in.; main plate, $\frac{5}{16}$in. Axles, 1¼in.; length, 48in. Wheels, 44in.; 1⅜in. spokes. Shafts: 2¼in. x 1½in.

No. 96.—RALLI CART. Body frame, width, 38in.; length, 56in.; depth of sides, 23in.; wings, 7in.; seat sills, length, 36in. Shafts, 2¼in x 1½in.; bar to points, 69in.; over all, 117in. Springs: over all, 50in.; 5 plates, 1½in. x ¼in.; main plates, $\frac{5}{16}$in. Axles: collinge, 1¼in.; over collars, 48in; Wheels, 48in.; spokes, 1½in.; hubs, 6in. x 8in.; rims, 1⅝in.

Working drawing appeared in *Coachbuilder*, Vol. 5, No. 5.

No. 97.—POLO CART. Body: length over bottom, including bracket, 54in; width across bottom, 35in.; top, 38½in.; depth of sides over top batten, 18in.; back bevel, 3in. Shafts, 2¼in. x 1½in. lancewood; bar to points, 71in. Springs: length over, 43in; 4 plates, 1½in. x ¼in. Axles: collinge, 1¼in.; over collars, 48in. Wheels, 52in.; spokes, 1½in.; hubs, 6in. x 8in.; rims, 1⅝in.

Working drawing appeared in *Coachbuilder*, Vol. 7, No. 11.

No. 98.—DOG-CART. Body: width—bottom, 32in.; top, 36in.; across upper panels, 39in.; length, bottom, including bracket, 62in.; upper panel, depth, 7in; length, 36in.; depth of side panels, 12in. Shafts, 2½in. x 1⅝in.; bar to points, 71in. Axles, 1⅜in.; length, 50in. Springs, 50in.; bolt, 2in. back from centre; 6 plates, 1½in. x ¼in.; cross spring, 38in.; 5 plates, 1½in. x ¼in. Wheels, 56in.; spokes, 1⅝in.; hubs, 6½in. x 8½in.

No. 99.—DOG-CART. Body: width—bottom, 32in.; top, 38in.; length over all, 60in.; depth over side, 18in. Shafts, 1⅝in. x 2⅝in.; bar to points, 71in. Springs, 48in.; 6 plates, 1½in. x ¼in.; main plate, $\frac{5}{16}$in.; compass inside to centres, 3½in. Axles, 1¼in.; length, 50in. Wheels, 54in.; 1⅝in. spokes.

No. 100.—RUSTIC CART, heavy size. Width—bottom, 38in.; top, 44in.; length over body, bottom, 58in.; seat sills, length, 38in. Wheels, 58in.; spokes, 1¾in.; hubs, 8in. x 10in.; rims, 2½in. Axles: mail, 1½in.; length, 50in. Shafts, 1¾in. x 2½in.; bar to points, 71in. Springs: side, 44in.; 6 plates, 1¾in. x ¼in.; main plate, $\frac{5}{16}$in.; compass, 6 in. Cross spring, 36in.; 4 plates, 1¾in.

No. 101.—RUSTIC CART, 2 seats. Body: width—bottom, 34in.; top, 40in.; length over bottom, 50in.; depth over sides, 18in.; bevel at back, 6in.; length of top batten, 34in. Shafts, 2½in. x 1⅝in.; bar to points, 72in. Springs, 46in.; 4 plates, 1¾in. x ¼in.; main plate, $\frac{5}{16}$in.; compass inside to centres, 4in. Axles: mail, 1¼in.; length, 48in. Wheels, 50in.; 1½in. spokes.

Working drawing appeared in *Coachbuilder*, Vol. 7, No. 4.

No. 102.—CURVED SIDED RUSTIC CART, 2 seats. Body: width—bottom, 29in.; over top battens, 45in.; depth of sides over all, 22in.; bevel back, ¾in.; length, bottom, 51in. Shafts, 2¼in. x 1¾in.; bar to points, 72in.; between shafts at bar, 32in. Springs, 45in.; 5 plates, 1½in. x ¼in.; main plate, $\frac{5}{16}$in.; compass inside, 4in. Axles, 1¼in.; length, 46in. Wheels, 51in.; spokes, 1½in.

Working drawing in *Coachbuilder*, Vol. 8, No. 1.

No. 103.—RUSTIC CART, 2 seats. Body: width—bottom, 30in.; top over battens, 38in.; depth over sides, 18in.; back, bevel, 5½in.; length, bottom, including bracket, 54in. Shafts, 2¼in. x 1¾in.; bar to points, 72in. Springs: side, 44in.; 5 plates, 1½in. x ¼in. bearing, 1in. back from centre; cross, 37in.; 4 plates, 1½in. x ¼in. Axles, 1¼in.; length, 48in. Wheels, 54in.; spokes, 1⅝in.

No. 104.—DOG-CART. Measurements same as in No. 98.

No. 105.—CURVED PANEL Pony Cart, 2 seats. Body: width under bottom panel, 30in.; top, 32in.; over curved panels, 44in.; length bottom over all, 56in.; depth of panels, 12in. Shafts, 1½in. x 2¼in.; bar to points, 67in. Springs, 44in.; 5 plates, 1½in. x ¼in. Axles, 1¼in.; length, 48in. Wheels, 46in.; spokes, 1⅜ in.

No. 106.—CURVED PANEL CART, 2 seats. Body: width—bottom, 34 in.; top of panel, 36 in.; over curved panels, 48in. Seats, 17 in. and 13 in.; length, bottom, 56 in.; depth of panels, 12 in. Shafts, 1½in. x 2¼in.; bar to points, 67 in. Springs, 42 in. over all; 4 plates 1½in. x ¼in. Axles, 1¼ in.; length, 48 in. Wheels, 48 in.; spokes, 1⅜in.

No. 107.—PONY POLO CART. Same as No. 97, excepting wheels, 48 in.; shafts, bar to points, 67 in.

Working drawing in *Coachbuilder*, Vol. 7 No. 11.

No. 108.—CURVED PANEL CART, 2 seats. Body same as No. 105, excepting depth of curved panels, 14 in. Springs: side, 46 in.; 5 plates, 1½ in. x ¼ in.; main plate, $\frac{5}{16}$ in.; compass inside to centres, 4in.; cross spring, 32in.; 4 plates, 1½in.; compass, 5in. Axles, 1¼in.; length, 48in. Wheels, 54in.; spokes, 1⅜in.

No. 109.—SPRING CART. Body: width, 40in.; length, bottom, 62in; depth of side over all, 21in.; length of seat sill, 40in. Shafts, 1¾in. x 2¾in.; bar to points, 72 in. Springs: side, 48 in.; 7 plates, 2in. x ¼in.; main plate, $\frac{5}{16}$in. Axles, 1¾in.; mail length, 50in., Wheels, 58in.; spokes, 2¼in.; hubs, 8½in. x 10½in.; felloes, 2¾ in.; tyres, 2in. x ¾in.

No. 110.—PAGNEL CART, 2 seats. Body: width—bottom outside, 38in.; top, 42in; length, bottom, including bracket, 61in.; depth of sides over rail, 20in.; bevel of back corner, 6½in. Wheels, etc., same as No 98.

No. 111.—GOVERNESS CAR. Body: width—under both seats, 42in.; over seats, 46in.; depth of seat panel, 9 in.; between seats, 17in.; width under bottom, 28in.; lengths, bottom, 33in.; under seats, 40in.; over seats, 44in. Shafts, 2in. x 1½in.; bar to points, 58in.; between at bar, 32in. Springs, 38in.; 4 plates 1½in. x ¼in.; compass inside, 8in. Axles, 1¼in.; between collars, 44in.; between cranks, 31in. Wheels, 44in.; spokes, 1¼in.

Working drawing appeared in *Coachbuilder*, Vol. 6, No. 4.

No. 112.—ALEXANDRA CART. Body: width—bottom, 33½in.; top, 36½in.; depth of body panels, 11½in.; depth of upper panel; 11in.; length over all, 59in. Seats, 15in. and 13in. Shafts, 2in. x 1⅝in. Lancewood, bar to points, 67in. Springs, 36in.; 4 plates, 1½in. x ¼in.; compass inside, 8in. Axle: 1¼in.; length, 48in. Wheels, 52in.; spokes, 1⅝in.

Working drawing appeared in *Coachbuilder*, Vol. 4, No. 7.

No. 113.—LENNOX GIG. Body: length—bottom, including bracket, 50in.; width, bottom, 28in. Seat, 18in. x 34in. Springs: side, length over all, 42in; 4 plates, 1½in. x ¼in. Axles: 1½in.; length, 46in. Wheels, 48in.; spokes, 1⅜in.

TABLE OF MEASUREMENTS.

No. 114.—CABRIOLET. BODY: width—bottom, 34in.; top of standing pillar, front, 44in.; back pillar, 38in.; depth of side over all, 24in. Shafts, 2⅜in. x 1⅜in.; bar to points, 74in. Springs: length over all, 54in.; 4 plates, 1½in. x ¼in.; main plate, $\frac{5}{16}$in. Wheels, 54in.; spokes, 1½in.; hubs, 5¾in. x 8in.; rims, 1⅝in.

Working drawing appeared in *Coachbuilder*, Vol. 10, No. 1.

No. 115.—SOUTHLAND ROAD CART. Body: width—bottom, 30in.; top, 34in.; depth of sides, 12in.; length, bottom, 42in. Seat, 36in. x 16in. Shafts, 2¼in. x 1⅜in.; bar to points, 67in. Springs: side, 42in.; 4 plates, 1½in. x ¼in.; compass inside to centres, 4½in.; cross, 34in.; 3 plates, 1½in. Wheels 52in.; spokes, 1⅜in.

No. 116.—HOODED ROAD CART. Body: width—across bottom, 30in.; top, 33in; length, bottom to dash, 44in. Seat, 36in. x 16in. Wheels, springs, etc., same as No. 115.

No. 117.—SPINDLE SEAT CART. Body: width—bottom, 30in.; top, 32in.; length over bracket, 44in. Seat, 16½in x. 34in.; depth over seat rails, 8in. Other measurements same as No 115.

No. 118.—DAISY CART (1 seat). Body: width—bottom, 29in.; top, 38½in.; length, bottom, over all, 51in.; depth of sides over all, 19in.; at back, 13½in. Shafts: 2¼in. x 1½in. Springs: 36in.; 4 plates 1½, x ¼in.; compass inside to centres, 8in. Axle: 1⅛in., 46in. Wheels, 50in.; 1⅜ sarven.

No. 119.—BRASSEY CART. Body: length on bottom, 44in.; width, bottom, 29in.; top, 33½in.; depth of side panels, 10in.; depth of seat risers, 3in.; seat on bottom, 37in. x 15in. Springs: side, 44in.; 4 plates, 1½in. x ¼in.; compass inside, 5in.; cross spring, 32in.; 3 plates, 1½in. x ¼in. Axle, 1¼in. h.p., length, 46in. Shafts, 2¼in. x 1½in.; bar to points, 74in. Wheels, 56in.; spokes, 1½in.

No 120.—VICTORIA ROAD CART. Body: width—bottom, 29in.; top, 32in.; depth of sides, 14in.; back, 9¼in.; leg room, 27in. Seat: 39 in. x 15½in. Shafts: 2¼in. x 1¾in.; bar to points, 74in.; between at bar, 34½in. Springs, 36in; 4 plates, 1½in. x ¼in.; compass inside, 8in. Axle, 1⅛in.; length, 50in. Wheels, 54in.; spokes, 1½in.;

Working drawing appeared in *Coachbuilder*, Vol. 10, No. 12.

No. 121.—GLADSTONE CART. Body: width—bottom, 26in; under curved panels, 34in.; over curved panels, 45in.; depth of under panels, 9½in.; top panels, 12in. Other measurements same as No. 115.

No. 122.—CURVED PANEL CART (2 seats). Body: width—bottom, 34in.; over top panels, 48in.; length over foot board, 59in.; depth of under panel, 12in.; upper, 13in. Seats, 16in. and 14in. Wheels, springs, etc., same as No. 118.

No. 123.—CUPID GIG. Body: width—bottom. 30in.; under seat, 33in.; depth of sides, 12in.; seat, 16in. x 36in. Wheels. springs, etc., same as No. 126.

No 124.—TAUIWHA CART (1 seat). Body: width—bottom, 32in.; length, bottom, 44in. Wheels, springs, etc., same as No. 126.

No. 125.—TARANAKI GIG. Body: width—bottom, 30in.; top, 33in.; length over all, 48in.; depth of sides over all, 12in. Seat, 16½in. x 36in. Wheels, 54in. Springs, etc., same as No. 115.

No. 126.—AUCKLAND ROAD CART. Seat: 36in. x 17in.: depth of end panels, 10½in. Tray, width, 29in. Shafts, 1½in. x 2½in.: bar to points, 80in.; between shafts at bar, 30in. Springs, side. 42in.; 4 plates, 1½in. x ¼in.; cross spring, 34in.; 3 plates 1¼in. x¼in. Axles. 1⅛in. length, 50in.; Wheels, 52in.; spokes 1⅜in.

No. 127.—CEE SPRING GIG. Body: bottom, 30in.; top 34in.; length, bottom, over bracket, 44in. Springs, length over all, 42in.; compass inside, 8in. Wheels, 52in.; spokes, 1⅜in. Shafts, 1½in. x 2¼in; bar to points, 67 in. Axles, 1⅛in.; length, 50in.

No. 128.—SPEEDING CART. Body: width—bottom, 30in. Seat, 34in. x 17in. Shafts, 1½in. x 2¼in.; bar to points, 75in.; between at bar, 33in. Springs, 48in. over all; 4 plates 1¼in. x ½in. Axles 1⅛in. h.p., length 48in. Wheels, 56in.; spokes, 1⅜in.

No. 129.—ROAD CART WITH CUPID WING BOARDS. Seat, 38in. x 16in. Tray, width, 31in.; length, 27in. Shafts, 1¾in. x 2½in.; bar to points, 80in.; between shafts at bar 32in. Wheels, springs, etc., same as No. 126.

No. 130.—CUPID ROAD CART. Same as No. 129 excepting springs,—elliptic, 36in.; 4 plates, 1½in. x ¼in.; compass inside, 8in.

No. 131.—PHAETON SULKY (with Spindle seat). Seat: 38in. x 16in. Shafts, 1⅝in. x 2½in.; over the bend, 10in.; width between at bar, 33in.; length, bar to points, 80in. Wheels, springs, etc., same as No. 126.

No. 132.—PHAETON SULKY (with solid seat). Seat, 17in. x 38in.: depth of end panel, 7in.; side flare, 2½in.; back do 3½in. Other measurements same as No. 131.

Working drawing appeared in the *Coachbuilder*, Vol. 6, No. 3.

No. 133.—BENT SHAFT SULKY. Same as No. 131.

No. 134.—STRAIGHT SHAFT SULKY. Seat, 38in. x 16in.; seat risers, 2in.; tray, drop from top of shafts, 12in. Shafts, 1¾in. x 2½in.; between at bar, 34in.; bar to points, 80in.; springs, 42in.; 4 plates 1¼in. x ¼in.; compass inside to centres, 4½in.; cross spring, 35in.; 3 plates, 1¼in. x ¼in.; compass 4in. Axles, 1⅛in.; length 52in. Wheels, 52in. sarven; spokes 1$\frac{5}{16}$in.

No. 135.—TRAY SULKY. Seat, 38in. x 16in.; above shaft, 6in. Shaft, 1⅝in. x 2½in; bar to points, 80in.; length over cross bars, 40in.; between at bars, 34in. Tray: length over all, 40in.; depth of sides, 5½in. Wheels, springs, etc., same as No. 134.

No. 136.—LINDSAY ROAD CART. Body: width—bottom, 26in. Seat, 36in. x 16in. Shafts, 10in. drop, size, 1½in. x 2½in.; bar to points, 67in.; between at bar, 30in. Wheels, springs, etc., same as No. 134.

No. 137.—BENT SHAFT SULKY. Measurements same as No. 131.

No. 138.—KING OF THE ROAD SULKY. Convertible for pole or shafts. Seat, 36in. x 16in.; above shafts, 8in. Framework, 2in. x 1½in.; length over cross bars, 57in.; width, 33in. No. of cross bars, 3. Shafts, 2¼in. x 1¾in.; length over all, 11ft 6in.; pole, 2½in. x 1¾in.; from front cross bar to point, 8ft. Springs: side, 48in.; 5 plates, 1½in. x ¼in.; compass inside, 5½in.; cross, 33in.; 4 plates, 1½in. x ¼in.; compass, 3½in. Axles, 1¼in.; length, 52in. Wheels, 54in.; sarven, 1⅜in.

Working drawing appeared in *Coachbuilder*. Vol. 7, No. 2.

No. 139.—Spindle Seat Sulky. Seat risers, depth, 5½in. Other measurements same as No. 134.

No. 140.—Straight Shaft Sulky (with solid seat). Depth of seat panels, 7in. Other measurements same as No. 134.

No. 141.—Road Cart. Same measurements as No. 142, excepting springs, 34in.; 4 plates, 1¼in. x 1¼in.; compass inside, 7in.

Working drawing appeared in the Coachbuilder, Vol. 8, No. 6.

No. 142.—Road Cart (on cee springs). Tray, width, bottom, 28in.; under seat, 30in.; depth over all, 13½in.; box under seat, depth, 7in.; leg room, 27in. Seat, 36in. x 15½in. Shafts, 2in. x 1½in.; bar to points, 66in.; width between at bar, 30in. Springs over all, 42in.; 4 plates, 1½in. x 1¼in.; compass inside, 7in. Axles: 1⅛in.; length, 46in. Wheels, 50in.; sarven, spokes, 1¼in.

No. 143.—Straight Shaft Sulky (with solid seat). Measurements same as No. 134.

No. 144.—Tray Sulky (with canvas boot). Seat, 38in. x 16in.; above shafts, 10in. Shafts, 1¾in. x 2½in.; bar to points, 80in.; over back and front cross bars, 56in; width between at bars, 30in. Tray, dropped below shafts, 2½in.; length over all, 52in. Spring, 48in.; 5 plates, 1½in. x ¼in.; compass, 5½in. Axles: 1¼in.; length, 52in. Wheels: 52in.; sarven, spokes, 1⅜in.

No. 145.—Straight Shaft Sulky (with spindle seat). Measurements same as No. 134.

No. 146.—Straight Shaft Sulky. Seat, 38in. x 16in.; seat risers, depth, 6in.; leg room, 30in. Shafts, 1¾in. x 2¼in.; bar to points, 68in.; between at bars, 32in. Springs: 36in.; 4 plates, 1½in. x ¼in.; compass inside, 8in. Axles: 1⅛in.; length, 52in. Wheels, 52in.; sarven, spokes, 1⅜in.

No. 147.—Butcher's Order Cart. Body: width—bottom, 39in.; top, 42in.; depth of side panels, 17½in. Name boards, length over all, 40in.; depth, 9in.; length of body, bottom, 50in.; foot board, 7½in. Shafts, 2¾in. x 1¾in. Springs, 50in.; 6 plates, 2in. x ¼in; compass inside to centre, 3in. Axles: mail, 1½in.; length, 50in. Wheels: 54in.; spokes, 2in.; hubs, 8in. x 10in.; felloes, 2½in. Tyres, 1¾in. x ¾in.

Working drawing appeared in *Coachbuilder*, Vol. 9, No. 8.

No. 148.—Milk Cart. Body: width—bottom outside, 38in.; top, 40in.; length over cross bars, 48in.; over footboard 56in.; depth of sides, 19in. Shafts, 2¾in. x 1¾in. Springs: side, 44in.; 6 plates, 1¾in. x ¼in.; main plate, $\frac{5}{16}$in.; compass inside to centres, 4½in.; cross spring, 32½in.; 4 plates, compass, 4in. Wheels, 54in.; spokes, 1¾in.; rims, 2¼in.; hubs, 7½in. x 9½in.; tyres 1½in. x ⅝in.

Working drawing in *Coachbuilder*, Vol. 8, No. 10.

No. 149.—Baker's Delivery Cart. Body: width—bottom, 38in.; top, 43in; length, bottom over cross bars, 52in.; length over bottom sides, 60in.; footboard, 17in.; depth of sides, 24in. Shafts, 3in. x 2in.; bar to points, 67in. Springs: side, 50in.; 8 plates, 2in. x ¼in.; main plate, $\frac{5}{16}$in.; compass inside to centres, 5in.; cross, 40in.; 5 leaves, 1¾in. x ¼in.; main plate, $\frac{5}{16}$in.; compass, 5in. Axles: mail, 1⅝in.; length, 50in. Wheels, 58in.; spokes, 2in.; felloes, 3in.; hubs, 8in. x 10in.

Working drawing in *Coachbuilder*, Vol 6, No. 5.

No. 150.—Spring Dray. Body: width—across bottom side pieces, 44in.; inside body, 46in.; length over cross pieces, 72in.; footboard, 9in.; shafts, 3in. x 2½in.; footboard to points, 75in. Springs, 50in.; six plates, 2¼in. x $\frac{5}{16}$in.; main plate, ⅜in. Axle: mail, 1⅝in. Wheels, 58in.; spokes, 2¼in.; felloes, 2¾in.; hubs, 8¾in. x 10½in.

Working drawing in the *Coachbuilder*, Vol. 7, No. 10.

No. 151.—Settler's Waggon, measurements same as No. 94.

No. 152.—Business Delivery Waggon. Body: width—bottom, 42in.; depth of sides over rails, 15in.; length over cross bars, 90in. Hind springs: side, 42in.; 6 plates, 1¾in. x $\frac{5}{16}$in.; cross, 40in.; 6 plates, 1¾in. x $\frac{5}{16}$ in. Front springs, 40in.; 5 plates, 1¾in. x $\frac{5}{16}$in. Axles, 1⅜in. and 1½in.; length, 52in. Wheels, 36in. and 50in.; spokes, 1⅝in.

No. 153 and 154.—Creamery Waggon. Body: width—outside, 48in.; inside, 46in.; length over all, 84in.; depth of sides, 15in.; under seat to bottom, 29in. Springs: side, 40in.; 6 plates, 2in. x ¼in.; main plate $\frac{5}{16}$in.; compass inside to centres, 4½in.; cross, 42in.; 8 plates, 2¼in. compass, 5in. Axles: 1½in.; length, 54in. Undercarriage, 3 reaches; length over, 60in. Wheels, 52in. and 44in.; spokes, 2in.

No. 155.—Market Waggon (light). Body: width—inside, 45in.; length inside, 75in.; depth of side boards, 12in.; over battens, 22in.; undercarriage 1 reach; length, 60in. Springs: side 34in.; 5 plates, 1¾in. x $\frac{5}{16}$in.; front, 34in.; 6 plates, 2in. x $\frac{5}{16}$in. Axles, 1⅜in.; length, 52in. Wheels, 50 and 44in.; spokes, 1⅝in.

Working drawing appeared in *Coachbuilder*, Vol. 7, No. 9.

No. 156.—Milk Waggon. Body: width—inside, 46in.; length inside, 93in. Undercarriage, 3 reaches; length, 60in. Springs: side, 36in.; 6 plates, 1¾in.x¼in.; main plate $\frac{5}{16}$in.; front, 36in.; 7 plates, 2in. x $\frac{5}{16}$in. Axles: front, 1⅜in.; hind, 1½in.; mail. length, 54in. Wheels: 50in. and 42in.; spokes, 1¾in.

No. 157.—Farmer's Spring Waggon. Body: width—bottom outside, 54in.; across floating rails, 70in.; length over cross bars, 8ft.; footboard, 8in.; depth of sides, back and front, 18in.; middle, 17in. Undercarriage: length, 45in.; width, 51in. Springs: hind, 42in.; 8 plates, 2¼in. x $\frac{5}{16}$in.; compass inside to centres, 5in.; front, 36in.; 6 plates, 2¼in. x $\frac{5}{16}$in.; compass, 4in. Axles: drabble, 1¾in. and 2in.; length, 54in. Wheels, 54in. and 36in., Warner; spokes, 2¼in.

Working drawing in the *Coachbuilder*, Vol. 10, No. 9.

No. 158.—Single Lorry. Tray: length over, 10ft.; width, 62in.; width over summers, 42in.; undercarriage, over bars, length, 48in.; width, 35in. Axle beds and bolsters, width, 5in. Wheels, in the wood, 35in. and 31in.: hubs, 8in. x 11in.; spokes, 2¼in.; felloes, 2¾in. Axles: drabble, 1¾in.; length, 40in. Springs: front, 36in.; 8 plates, 2¼in. x $\frac{5}{16}$in.; compass inside, 6½in.; hind, 38in.; 9 plates, 2¼in. x $\frac{5}{16}$in.; compass, 8in.; checkspring, 4 plates, 2¼in. x $\frac{5}{16}$in.

Working drawing in *Coachbuilder*, Vol. 6, No. 3.

No. 159.—Table Top Farm Waggon. Body. width—over summers, 52in.; body over all, 78in.; length of summers over cross bars, 13ft. Summers got out of 10in. x 2½in. Sweep on top of summers 2in.; width of bolsters and axle beds, 7in. Axles, 2¾in.; length, 59in. Wheels, 54in. and 44in.; spokes, 3½in.

No. 160.—Table Top Waggon. Body: widths—across summers, 50in.; over side pieces, 85in.; length over cross pieces, 18ft.; over top rails, 19ft. 8in. Thickness of summers, 3in.; sweep of summers, 2in.; side pieces, dressed to 4in. x 3in.; hind axle bed and bolster, 7in. deep x 8in. wide; front axle bed, 7in. x 7in.; bolsters, 5in. and 4in. x 7in. Undercarriage: length, 70in.; width over hounds, front, 50in.; back 45in. Axles, 3¼in.; length, 58½in.; length of box, 12in. Shafts, length over all, 8ft. 6in.; width, 36½in. Wheels, in the wood, 72in. and 56in.; spokes, 5in. x 1⅝in.; felloes, 5in. x 4½in.; 14 and 12 spokes.

Working drawing in *Coachbuilder*, Vol. 10, No. 7.

GENERAL INFORMATION

TO DRAW AN ELLIPSE OF GIVEN DIMENSIONS.

The major axis is the long diameter. The minor axis is the short diameter.

Let major and minor axis be AB and CD respectively. Make these lines to bisect each other at right angles at E. Open the compass

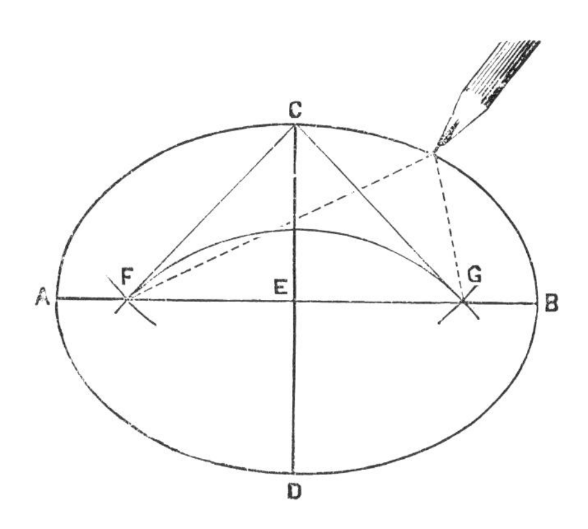

half the distance between A and B, that is, from E to B or A. From C as a centre make a circle with radius EB cutting AB in F and G. Put pins in at the points F C and G. Tie a loop tightly round the pins. Remove the pin at C, and then with a pencil in the loop trace the oval ADCB, which will be of the given dimensions.

TO OBTAIN NATURAL CURVES.

To obtain a natural curve for a spring. Say it is required to make a curve for a spring 42 inches long from centre to centre and 3 inches dip. Mark off on a board three points, A B and C, C being the thickness of the steel more than the dip required. Drive stout nails at A and B, put a piece

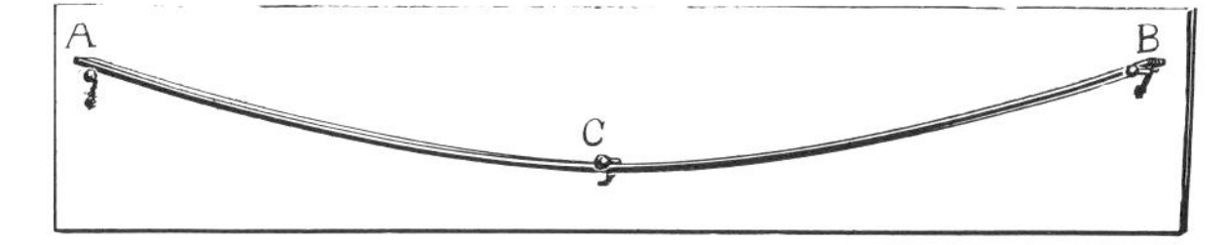

of spring steel across. Then spring the steel back to C, and nail it there. The curve made by the steel will be a natural curve, which, if given to the main plate, will make the spring straight when deflected the amount of the dip.

Another way to get a natural curve is to take a board, and put pins in at D and E, and mark the point F, which determines the drop of the curve. A piece of silk, a supple chalk line suspended between D and E. The curve made by the string or thread will be a natural flowing curve, suitable for many lines in vehicle construction.

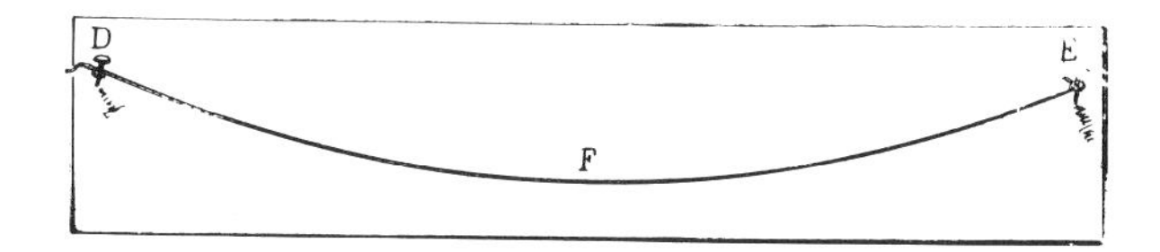

To get more swell one end or the other raise one of the pins. If the line is chalked and the board held so that it will fall against it, and the board gently turned over on the flat by placing another board gently on top of the chalked line, the curve can be transferred to the board.

A piece of thin spring steel, say $\frac{1}{2} \times \frac{1}{16}$ is very useful for marking spirals, curves, and graceful lines, in which the most artistic eye will find little to complain of.

ESTIMATED CARRYING CAPACITY OF AXLES.

These sizes are approximate only, and are for axles of ordinary lengths between collars and ordinary heights of wheels. Manufacturers generally recommend the use of axles larger than the size which will just carry the load.

AMERICAN AXLES, PER SET OF FOUR ARMS.

Front and Back.	Capacity.
$1\frac{1}{8}$ in.	800 Lbs.
$1\frac{1}{4}$ in.	1,000 ,,
$1\frac{3}{8}$ in.	1,500 ,,
$1\frac{1}{2}$ in.	2,000 ,,
$1\frac{5}{8}$ in.	3,000 ,,
$1\frac{3}{4}$ in.	4,000 ,,
$1\frac{7}{8}$ in.	5,000 ,,
2 in.	6,000 ,,
$2\frac{1}{8}$ in.	7,000 ,,
$2\frac{1}{4}$ in.	8,000 ,,
$2\frac{1}{2}$ in.	9,000 ,,
$2\frac{3}{4}$ in.	10,000 ,,
3 in.	12,000 ,,

ENGLISH AXLES.

DRAY AXLES, PER PAIR ARMS.

$1\frac{3}{4}$ in.	1,700 Lbs.
2 in.	2,200 ,,
$2\frac{1}{4}$ in.	2,800 ,,
$2\frac{1}{2}$ in.	3,300 ,,
$2\frac{3}{4}$ in.	5,000 ,,
3 in.	6,100 ,,

ESTIMATED CARRYING CAPACITY OF AXLES—*Continued.*

MAIL PATENT OR COLLINGE PATTERNS,

PER SET OF FOUR ARMS.

1 in.	600 Lbs.
$1\frac{1}{8}$ in.	800 ,,
$1\frac{1}{4}$ in.	1,100 ,,
$1\frac{3}{8}$ in.	1,500 ,,
$1\frac{1}{2}$ in.	1,900 ,,
$1\frac{5}{8}$ in.	2,400 ,,
$1\frac{3}{4}$ in.	3,000 ,,
$1\frac{7}{8}$ in.	3,600 ,,
2 in.	4,400 ,,

DRABBLE AXLES, PER SET OF FOUR ARMS.

$1\frac{1}{4}$ in.	1,100 Lbs.
$1\frac{1}{2}$ in.	2,000 ,,
$1\frac{5}{8}$ in.	2,400 ,,
$1\frac{3}{4}$ in.	3,700 ,,
$1\frac{7}{8}$ in.	4,400 ,,
2 in.	4,900 ,,
$2\frac{1}{4}$ in.	6,100 ,,
$2\frac{1}{2}$ in.	7,400 ,,
$2\frac{3}{4}$ in.	10,000 ,,
3 in.	13,000 ,,

THE STRENGTH OF SPRINGS.

It is not possible to construct a table to show carrying capacities of springs which shall be true for all springs of the same dimensions.

The strength and resilience of springs depend not only upon the quality of the steel, but upon its dimensions, and the treatment it receives at the hands of the spring maker. The difference in qualities of steel alone make a universal table of carrying capacities impossible. The difference in the methods of treating steel magnify the difficulty.

It will therefore be understood that no claim of infallibility is made for the following tables.

The laws which govern the action of springs, however, have been ascertained, and with a fair degree of accuracy expressed in mathematical terms. The quality and temper of the steel determine the expressions which are known as "Proof Stress on the most Fatigued Fibre," or "Proof Stress" and "Modulus of Elasticity."

The "Proof Stress" is the amount of stress per square inch of section which a piece of steel will withstand without permanently changing its shape. It varies from about 30,000 lbs. per square inch of section in ordinary Bessemer steel up to 100,000 lbs. in good spring steel.

The Modulus of Elasticity is fairly constant for all grades of iron and steel, and is generally taken at about 30,000,000 lbs.

The tables which follow are calculated for steel with a proof stress of 80,000 lbs. per square inch, and a Modulus of Elasticity of 30,000,000 lbs.

They are calculated from the formula given in most works on applied mechanics, namely

$$W = \frac{n\ b\ t^2 f.}{6\ l}$$

In the above formula $W = \frac{1}{2}$ load producing proof stress, n = number of plates, b = width of plate, t = thickness of plate, f = proof stress, and $l = \frac{1}{2}$ developed length of main plate.

Another way of putting the above is to say the load which will subject either half the main plate to its proof stress is—supposing all plates the same thickness—the number of plates multiplied by the area of section of one plate, and by its thickness and the number which represents Proof Stress, and their combined product divided by half the length of the main plate multiplied by 6, the quotient will be the proof load for half the spring. That is, Proof Load for half of the spring = number of plates × area of section of one plate × thickness of one plate × number representing Proof Stress, and their combined product ÷ half the length of the main plate × 6.

The proof load for half the spring is taken as safe load for the whole spring. The values for $\frac{1}{2}$W given in the tables represent the safe load for a single plate. To ascertain a safe load for a complete spring.

Multiply the number in the column under the figures representing the given width of plate, and opposite the required length, by the number of plates, and the product may be taken as the safe load expressed in pounds, for the spring.

Example: To ascertain safe load for a spring 42 x $1\frac{1}{2}$ x $\frac{1}{4}$x4 plates, take from table for $\frac{1}{4}$ in. plates number opposite 42 and in the column under $1\frac{1}{2}$ in., that is 60. Multiply 60 by 4, the number of plates.

The product 240 represents in lbs. the safe load for 1 spring.

For quick moving vehicles the safe load will be less, and for slow moving vehicles, more, than the values ascertained from the tables. The load includes weight of body of vehicle.

Left Hand Columns show length of main plate. Top line width of plates, and any number in columns shows safe load in pounds for one plate of the length opposite it in left hand, and width above it in top line.

LENGTH OF MAIN PLATE	STEEL 3/16-INCH THICK.						
(2*l*) = in.	1″	1¼″	1½″	1¾″	2″	2¼″	2½″
30 ,,	31¼	39	47	54	62	70	78
32 ,,	29¼	36	44	50	58	65	72
34 ,,	27½	34	41	48	55	61	68
36 ,,	26	32	39	45	52	58	64
38 ,,	24½	30	37	42	49	54	60
40 ,,	23½	29	35	41	47	52	58
42 ,,	22¼	27½	33	39	44	49	55
44 ,,	21¼	26	32	37	42	47	52
46 ,,	20½	25	31	35	41	45	50
48 ,,	19½	24	30	34	39	43	48
50 ,,	18¾	23½	28½	33	37	42	47
52 ,,	18	22½	27	31	36	40	45
54 ,,	17½	21½	26	30	35	39	43
56 ,,	16¾	21	25	29	34	38	42

LENGTH OF MAIN PLATE	STEEL 9/32NDS. INCH THICK.						
(2*l*) = in	1″	1¼″	1½″	1¾″	2″	2¼″	2½″
30 ,,	70¼	87	105	122	140	157	174
32 ,,	65¼	81	98	114	130	146	162
34 ,,	62	77	93	108	123	139	154
36 ,,	58½	73	87	102	117	131	146
38 ,,	55½	69	83	97	111	124	138
40 ,,	52¾	65	79	91	105	117	130
42 ,,	50	62	75	87	100	112	124
44 ,,	48	60	72	84	96	108	120
46 ,,	46	57	69	80	92	103	114
48 ,,	44	55	66	77	88	99	110
50 ,,	42	52	63	73	84	94	104
52 ,,	40½	50	60	70	81	90	100
54 ,,	39	49	58	68	78	87	98
56 ,,	37½	48	56	67	75	85	96

	STEEL 7/32NDS. INCH THICK.						
	1″	1¼″	1½″	1¾″	2″	2¼″	2½″
30 ,,	42½	53	63	74	85	95	106
32 ,,	40	50	60	70	80	90	100
34 ,,	37½	47	56	66	75	85	94
36 ,,	35½	44	53	62	71	80	86
38 ,,	33½	42	50	58	67	75	84
40 ,,	32	40	48	56	64	71	80
42 ,,	30¼	38	45	53	60	68	76
44 ,,	29	36	43	50	58	65	72
46 ,,	27¾	35	41	48	55	62	70
48 ,,	26½	33	39	46	53	59	66
50 ,,	25¼	31	37	44	50	56	62
52 ,,	24½	30	36	42	49	54	60
54 ,,	23½	29	35	41	47	52	58
56 ,,	22¾	28	34	40	45	50	56

	STEEL I = 5/16INCH THICK.						
	1″	1¼″	1½″	1¾″	2″	2¼″	2½″
30 ,,	86¾	108	129	151	173	194	216
32 ,,	81½	101	123	142	163	182	202
34 ,,	76½	95	114	133	153	171	190
36 ,,	72½	90	108	126	145	162	180
38 ,,	69	86	103	120	138	155	172
40 ,,	65	81	98	114	130	146	162
42 ,,	62	77	93	108	124	139	154
44 ,,	59½	74	89	104	119	133	148
46 ,,	57	71	85	99	114	128	142
48 ,,	54¼	68	81	95	108	122	136
50 ,,	52¼	65	78	91	104	117	130
52 ,,	50¼	62	75	87	100	112	124
54 ,,	48¼	60	72	84	96	108	120
56 ,,	46½	58	69	81	92	104	116

	STEEL ¼-INCH THICK.						
	1″	1¼″	1½″	1¾″	2	2¼″	2½″
30 ,,	55½	69	83	97	107	124	138
32 ,,	52	65	78	91	104	117	130
34 ,,	49	61	73	85	98	110	122
36 ,,	46¼	58	69	81	92	104	116
38 ,,	44	55	66	77	88	99	110
40 ,,	41¾	52	63	74	83	94	104
42 ,,	39¾	50	60	70	79	90	99
44 ,,	37¾	47	56	66	75	85	94
46 ,,	36¼	45	54	63	72	81	90
48 ,,	34¾	43	52	60	69	78	86
50 ,,	33¼	41	50	57	66	74	82
52 ,,	32	40	48	56	64	72	80
54 ,,	31	39	46	54	62	70	78
56 ,,	29¾	37	45	52	59	68	75

	STEEL I = 11/32INCH THICK.						
	1″	1¼″	1½″	1¾″	2″	2¼″	2½″
30 ,,	105	131	155	183	210	236	262
32 ,,	98½	122	147	170	197	220	244
34 ,,	92½	115	138	161	185	207	230
36 ,,	87½	109	131	153	175	196	218
38 ,,	82½	103	123	145	165	195	206
40 ,,	78½	98	117	137	157	176	196
42 ,,	75	94	112	131	150	169	188
44 ,,	71½	89	107	125	143	160	178
46 ,,	68½	85	102	119	137	152	170
48 ,,	65½	81	98	114	131	146	163
50 ,,	63	79	95	110	126	142	157
52 ,,	60½	75	91	105	121	135	150
54 ,,	58½	72	87	101	117	130	144
56 ,,	56	70	84	98	112	126	140

	STEEL ⅜INCH THICK.						
	1″	1¼″	1½″	1¾″	2″	2¼″	2½″
30 ,,	125	156	187	218	250	281	312
32 ,,	117	146	175	204	234	263	292
34 ,,	110½	137	165	192	221	247	274
36 ,,	104	130	156	182	208	234	260
38 ,,	99	124	149	173	198	223	248
40 ,,	94	117	141	165	188	211	234
42 ,,	89½	112	133	157	179	201	224

	STEEL ⅜INCH THICK.						
	1″	1¼″	1½″	1¾″	2″	2¼″	2½″
44 ,,	85	106	128	150	170	191	212
46 ,,	81½	102	122	143	163	183	204
48 ,,	78	97	117	136	156	175	194
50 ,,	75	94	113	131	150	169	188
52 ,,	72	90	108	126	144	162	180
54 ,,	69½	87	104	122	139	156	174
56 ,,	67	84	101	118	134	151	168

WEIGHT, STRENGTH, AND ELASTICITY OF TIMBERS, PRINCIPALLY AUSTRALIAN.

NEW SOUTH WALES.

TIMBER	WEIGHT	MODULUS OF RUPTURE	MODULUS OF ELASTICITY
Tallow Wood ...	77	15,300	2,288,000
Spotted Gum ...	62	13,300	2,056,000
Black Butt ...	67	13,700	2,163,000
Swamp Mahogany	76	12,100	2,099,000
Grey Ironbark ...	74	17,900	2,485,000
Red Ironbark ...	77	16,300	2,342,000
White I onbark ...	74	16,900	2,794,000
Wooly Butt ...	64	12.700	2,140,000
Red Gum	62	6,900	762,000
Grey Gum	57	13,000	2,147,000
Flooded Gum ...	69–77	14,800	2,140,000
Mountain Ash ...	66	11,500	2,054,000
Blackwood... ...	37–45	10,300	1,908,000
Grey Box	74	16,200	2,760,000
Pine	54	8,800	2,408,000
Forest Mahogany...	72	13.800	3,041,000
Rosewood	74	10,600	1,937,000
White Beech ...	63	15,600	2,421,000
Mahogany	75	14.500	2,258,000
Forest Oak ...	75	15,500	2,396,000
Turpentine ...	69	11,700	1,965,000
Stringy Bark ...	71	13,900	2,353,000
Australian Teak ...	63	14,400	2,174,000

VICTORIAN.

TIMBER	WEIGHT per cubic ft	MODULUS OF RUPTURE	MODULUS OF ELASTICITY
Blue Gum (Vic. and Tas.)	60	13,000	1,747,000
Mountain Ash (Vic.)	44	13,700	2,026,000
Red Gum	66	11,200	1,340,000
Blackwood ...	43	10,500	1,286,000
Mahogany	60	12,100	1,700.000

SOUTH AUSTRALIAN.

TIMBER	WEIGHT per cubic ft.	MODULUS OF RUPTURE	MODULUS OF ELASTICITY
Sugar Gum ...	70	10,200	1,620,000
Box Gum	72	12,900	2,549,000
Blue Gum	66	12,600	1,931,000
Red Gum	59	6,600	900,000

WEST AUSTRALIA.

TIMBER	WEIGHT per cubic ft.	MODULUS OF RUPTURE	MODULUS OF ELASTICITY
Jarrah	56 to 67	12,000	1,800,000
Red Gum (Like Jarrah ...	66	12,000	2,100,000
Karri	59	9,600	1,707,000

QUEENSLAND.

TIMBER	WEIGHT per cubic ft.	MODULUS OF RUPTURE	MODULUS OF ELASTICITY
Iron Bark	74	15,000	2.300,000
Spotted Gum ...	72	17,000	2,500,000
Tallow Wood ...	71	10,600	1,590,000
Blood Wood ...	56	9,000	1,200,000
Turpentine ...	70	19,300	2,800,000

EUROPEAN AND OTHER FOREIGN TIMBERS.

TIMBER	WEIGHT (Per Cubic Foot).	MODULUS OF RUPTURE
Ash	47	13,000
Beech	44	10,500
Birch	44	11,700
Bullet-tree	63	19,000
Chestnut	33	10,660
Ebony	75	27,000
Elm	33	7,900
Fir, Red Pine	31 to 44	8,300
,, Spruce	31 to 44	11,100
,, Larch	31 to 34	7,500
Greenheart	63	22,000
Kauri Pine (N.Z.) ...	36	11,000
Lancewood	42 to 62	17,350
Lignum Vitæ	40 to 83	12,000
Mahogany (Honduras) ...	34	11,500
,, (Spanish) ...	52	7,600
Oak (British)	44 to 62	11,800
,, (Dantzic)	44 to 62	8,700
,, (American)	62	10,600
Sycamore	37	9,600
Teak (Indian)	40 to 53	15,500
Willow	25	6,600

AMERICAN TIMBERS.

TIMBER.	MODULUS OF RUPTURE.	MODULUS OF ELASTICITY.	ELASTIC LIMIT.	WEIGHT PER CUBIC FOOT IN LBS.
Shagbark Hickory	16,000	2,390,000	11,200	51
Mockernut ,,	15,200	2,320,000	11,700	53
Water ,,	12,500	2,080,000	9,800	46
Bitternut ,,	15,000	2,280,000	11,100	48
Nutmeg ,,	12,500	1,940,000	9,300	49
Pecan	15,300	2,530,000	11,600	49
Pignut	18,700	2,730,000	12,600	56
White Elm ...	10,300	1,540,000	7,300	34
Cedar ,, ...	13,500	1,700,000	8,000	46
White Ash ...	10,800	1,640,000	7,900	39
Green ,, ...	11,600	2,050,000	8,900	39

It may be taken as a rule, that, given two pieces of wood of the same kind, both seasoned equally, the heavier piece is the stronger.

For instance, of two pieces of seasoned hickory choose the heavier for strength.

The Modulus of rupture is eighteen times the load which is required to break a bar 1 in. square supported at two points 1 foot apart, and loaded in the middle between the points of support, or

$$S = \frac{3\,w\,l}{2\,b\,d^2}$$

When W = load applied at middle between supports, *b* represents breadth, *d* represents depth, *l* represents length of beam, and S the modulus of rupture, or breaking stress per square inch.

The amount of extension or deflection of a piece of elastic material under stress is proportionate to stress.

The Modulus of Elasticity represents an imaginary fact, which can only be realised in such a material as Indiarubber ; it represents the weight which would double the length of a piece of material.

RELATIVE STRENGTHS OF BEAMS DIFFERENTLY FIXED AND LOADED.

For beams of the same material and of equal length, breadth, and depth, the following is true :—

(A). If a beam supported at the ends and loaded in the middle will carry 100 lbs. and deflect 1 inch

(B). a beam with load uniformly over its whole length will carry 200 lbs. and deflect $\frac{5}{8}$ of an inch

(C). a beam with ends fixed and loaded in the middle will carry 200 lbs. and deflect $\frac{1}{4}$ inch

(D). a beam fixed at ends with load uniformly spread over its length will carry 300 lbs and deflect $\frac{1}{8}$ in.

(E). a beam fixed at one end and loaded at the other will carry 25 lbs. and deflect 16 inches

(F). a beam fixed at one end and load uniformly spread over its whole length will carry 50 lbs. and deflect 6 inches.

Taking A as a standard the relative strengths and deflections are as follows:—

A	strength	1	deflections	1
B	,,	2	,,	$\frac{5}{8}$
C	,,	2	,,	$\frac{1}{4}$
D	,,	3	,,	$\frac{1}{8}$
E	,,	$\frac{1}{4}$	,,	16
F	,,	$\frac{1}{2}$	,,	6

In general the strength of a beam is represented by the Modulus of Rupture for the given material multiplied by area of section, and by depth of beam, and divided by length between supports or

$$\text{Strength} = \frac{s\,b\,d^2}{l}$$

That is strength equals area of section multiplied by depth and by the number given in tables as Modulus of Rupture, and the product divided by length, all measurements in inches. In the formula as expressed in symbols S represents the Modulus of Rupture taken from the table above, *b* the breadth of the beam, *d* the depth, and *l* the length.

When a factor of safety is introduced, the product should be further divided by the number representing factor of safety. The same units of measurement should be used throughout. If breadth and depth are expressed in inches, length also must be expressed in inches.

Of beams of rectangular sections the law is: The strength of a beam is doubled if we double its breadth or half its length. If we double its thickness we increase its strength four times.

FACTORS OF SAFETY.

MATERIAL	A DEAD LOAD OR ONE THAT DOES NOT ALTER	A LIVE LOAD OR ONE THAT ALTERS		IN STRUCTURES SUBJECTED TO SHOCKS
		IN TEMPORARY STRUCTURES	IN PERMANENT STRUCTURES	
Wrought Iron & Steel	3	4	4 to 5	10
Cast Iron ...	3	4	5	10
Timber ...	—	4	10	
Brick Work	—	—	6	
Masonry ...	20	—	20 to 30	

APPROXIMATE WEIGHT OF TYRE STEEL OR ROUND EDGE IRON.

PER SET OF 52 FEET.

Width	3-16 in.	1/4 in.	5-16 in.	3/8 in.	7-16 in.	1/2 in.	9-16 in.	5/8 in	3/4 in.	7/8 in.	1 in.
1 in.	35	48	58	71							
1 1/8 in.	39	53	65	80							
1 1/4 in.	44	63	73	88	104						
1 3/8 in		65	82	99	113	129					
1 1/2 in.			91	106	123	145	158	183			
1 5/8 in.				114	133	156	171	195	228		
1 3/4 in.				123	144	165	187	205	249		
2 in.				140	164	187	210	232	281	327	375

RULES FOR ORDERING SPRINGS.

In ordering springs a definite rule should be observed for measuring all springs of ordinary pattern. The generally accepted rule with English and Colonial spring makers is, for length, to measure from centre to centre of eyes; and for compass, from inside to a line passing through centre of eyes, as per the dotted lines in the following sketches:—

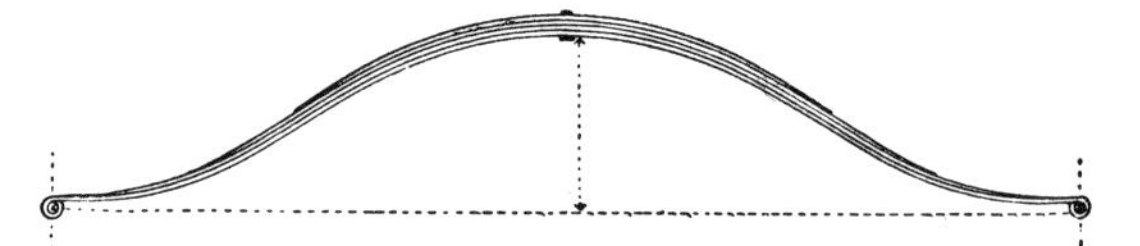

Fig. 1. Side springs; compass; inside the main plate to a line through both eyes; length from centre to centre of eyes.

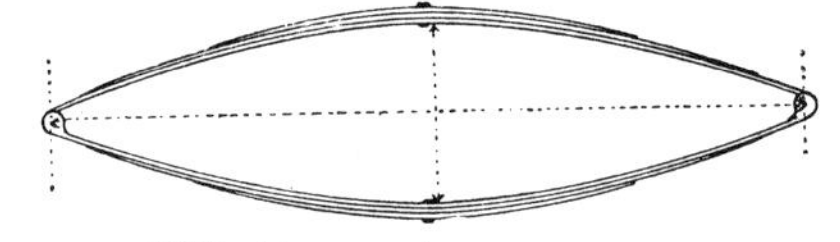

Fig. 2. Elliptic springs; compass; opening between main plates at middle; length, from centre to centre of eyes.

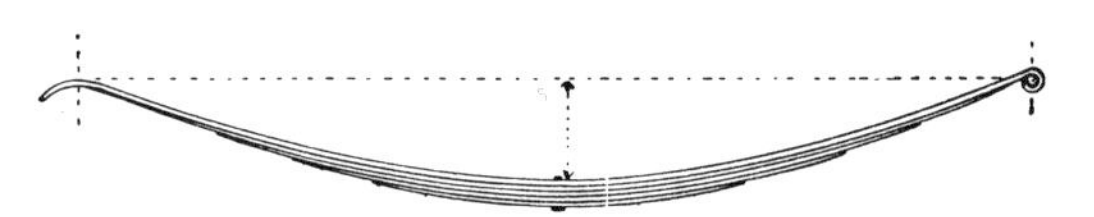

Fig. 3. Slipper springs; compass; opening between inside of main plate and a line through centre of one eye and touching the slipper bearing; length from centre of eye to centre of slipper bearing.

CIRCUMFERENCE OF CIRCLES.

Diameter in feet shown in left hand column, and circumference under second column. For circumference of——feet plus——inches, take figure in line opposite the number of in. first column and under the number of inches shown on top line. Example: Circumference of a circle 3 ft. 6 ins. in diameter is shown in line opposite 3 ft. under 6 in., that is 10 ft. 11 7/8 in. In measuring iron for tyres when inside diameter is given, add to circumference as given in the table three times the thickness of the iron. For measuring circle plates for a given inner diameter, add three times the width of the iron to circumference given in the tables. If outer diameter is given, subtract three times the width of the iron.

	0		1 in.		2 ins.		3 ins.		4 ins.		5 ins.	
ft.	ft.	ins.	ft.	ins.	ft.	ins.	ft.	ins.	ft.	ins.	ft.	ins.
1	3	1 5/8	3	4 5/8	3	8	3	11	4	2 1/8	4	5 3/8
2	6	3 3/8	6	6 1/2	6	9 5/8	7	0 3/4	7	3 7/8	7	7
3	9	5	9	8 1/4	9	11 3/8	10	2 1/2	10	5 5/8	10	8 3/4
4	12	6 3/4	12	9 7/8	13	1	13	4 1/8	13	7 1/4	13	10 1/2
5	15	8 1/2	15	11 5/8	16	2 3/4	16	5 3/4	16	9	17	0 1/8
6	18	10 1/8	19	1 1/4	19	4 3/8	19	7 1/2	19	10 5/8	20	1 7/8
7	21	11 7/8	22	3	22	6 1/8	22	9 1/4	23	0 3/8	23	2 1/8
8	25	1 1/2	25	4 5/8	25	7 7/8	25	11	26	2 1/8	26	5 1/4

	6 ins.		7 ins.		8 ins.		9 ins.		10 ins.		11 ins.	
ft.	ft.	ins.	ft.	ins.	ft.	ins.	ft.	ins.	ft.	ins.	ft.	ins.
1	4	8 1/2	4	11 5/8	5	2 3/4	5	5 7/8	5	9	6	2 1/4
2	7	10 1/4	8	1 3/8	8	4 1/2	8	7 5/8	8	10 3/4	9	1 7/8
3	10	11 7/8	11	3	11	6 1/8	11	9 3/8	12	1 1/2	12	3 5/8
4	14	1 5/8	14	4 5/8	14	7 7/8	14	11	15	2 1/8	15	5 1/4
5	17	3 1/4	17	6 3/8	17	9 5/8	18	0 3/4	18	3 7/8	18	7 1/8
6	20	4 7/8	20	8 1/8	20	11 1/2	21	2 3/8	21	5 1/2	21	8 3/4
7	23	6 3/4	23	11	24	1 1/8	24	4 1/8	24	7 1/4	24	10 3/8
8	26	8 3/8	26	11 1/2	27	2 3/4	27	5 3/4	27	9	28	0 1/8

GENERAL MEASUREMENTS OF ENGLISH CARRIAGE BODIES.

After CHRIS. W. TERRY, of London School of Carriage Drafting

	VEHICLE.	LENGTH ON ELBOW LINE.	WIDTH ON SEAT RAIL BETWEEN PILLARS.	HEIGHT FROM GROUND.	HEAD ROOM	SIZE OF EDGE PLATES.
		INCHES	INCHES	INCHES.	INCHES.	INCHES.
1	Four Horse Coach	56 to 62	39 to 42	32	42	1¾ by ⅜
2	Private Omnibus, to seat 6 inside	60	49	24	46	3 by ⅜, edge plate to boot
3	Pair-horse Landau	74	45	24	44	3 by ⅝ to 3 by ½
4	One-horse Landau	64	40	22	42	3 by ¾ to 3 by ⅝
5	Pair-horse Barouche	74	45	28	44	3 by ½ to 3 by ⅜
6	One-horse Barouche	66	42	23	43	3 by ⅝ to 3 by ½
7	" Barker " Brougham	50	42	21	42	1¾ by ⅜ body to 2 by ⅜
8	Double Brougham	48	45	23	45	1¾ by ⅜ body to 2¼ ,,
9	Double Brougham...	51	44	22	44	1¾ by ⅜ body to 2¼ ,,
10	Miniature Broughams	44	39	21	42	1¾ by 5/16 body to 2 by 5/16
11	Sociables	66 to 72	42 to 44	22 to 24	44	2½ by ⅝ to 2½ by ½
12	Phaetons— Victorias		39 to 44	21 to 24	42 to 44	2½ to 2½ by ⅜
13	Mail		40 to 42	36 to 37	46	2½ by ⅜ arch plate
14	Stanhope		38	34	45	2 by ¼ arch plate
15	Pony, Park		39	26		

	VEHICLE.	DIAMETER OF WHEELS. FRONT.	HIND	AXLES LENGTH OVER COLLARS	BETWEEN SPRING AND COLLAR.	CEE SPRINGS, HEIGHT. FRONT.	HIND.	ELLIPTIC SPRINGS, LENGTH. FRONT.	HIND.
		INCHES	INCHES	INCHES	INCHES	INCHES.	INCHES.	INCHES.	INCHES.
1	Drag or Four Horse Coach	38	49	51	—	side springs 28	side 28	Cross Spr'ng 44	——
2	Private Omnibus to carry 6 passengers...	38	47	52	2¼	——	——	38	38
3	Pair-horse Landau	38	46	48	2¾	27	29	37	under S. 38
4	Single-horse Landau ...	36	44	46	2	——	——	36	36
5	Pair-horse Barouche ...	38	46	47	2¼	26	28	37	37
6	One-horse Barouche ...	37	44	45	2	——	——	36	40
7	" Barker " Brougham ...	38	46	50	2	18	25	36	36
8	Double Brougham	38	46	50	2¼	18	24	36	36
9	Double Brougham	37	45	49	2	——	——	38	38
10	Miniature Brougham ...	36	41	47	2	——	——	36	37
11	Sociable	38	45	49	2	——	——	37	39
12	Phaetons—Victoria ...	38	45	48	2	——	——	36	39
13	Mail	35	42	49	—	side springs	——	36	34
14	Stanhope	34	41	46	—	——	——	36	38
15	Pony	34	40	46	—	——	——	36	37

LENGTH OF SHAFTS, ETC., FOR ENGLISH TWO-WHEELED VEHICLES.

	HEIGHT OF HORSES IN HANDS.				
	16	15	14	13	12
	ft. in.	ft. in.	ft. in.	ft. in.	ft in.
Length of Shafts	6 4	5 11½	5 7	5 2½	4 10
Height at tug to under side of shaft	4 2	3 11	3 8	3 5	3 2
Distance from tug to bar ..	4 4	4 1	3 10	3 7	3 4
Breeching staple to bar ..	2 5	2 3	2 1½	2 0	1 10½

LENGTH OF POLES, ETC., FOR PAIR-HORSE VEHICLES.

	HEIGHT OF HORSES IN HANDS.				
	16	15	14	13	12
	ft. in.	ft. in.	ft. in.	ft. in.	ft. in.
Length of Pole	9 0	8 6	8 0	7 6	7 0
Length of Splinter Bar ..	5 6	5 2	4 10	4 7	4 4
Height of Pole	3 8	3 5	3 2	2 11	2 8

NOTE.—For ponies it is sometimes necessary to make a Splinter Bar longer than given above.

Length of Leading Bars for Horses :—

Main Bar, 39 inches. Swingle Bars, each 33 inches over all.

FOR PONIES :—

Main Bars, 36½ inches. Swingle Bars, 31 inches over all.

COLOURS IN TEMPERING STEEL.

		DEG. FAHR.
Very pale straw	For Tools and Cutters for Metal	430
A shade of darker yellow		450
Darker straw yellow ..	For Edge Tools, Taps, etc.	470
Still darker straw yellow		490
Brown yellow	For Saws, Hatchets, etc ..	500
Yellow, tinged with purple		520
Light purple		530
Dark purple	For Springs	550
Dark blue		570
Paler blue	too soft for the above purposes	590
Still paler blue ..	,, ,, ,, ,,	610
Still paler blue with a tinge of green	,, ,, ,, ,,	630

SHRINKAGE and LOSS OF WEIGHT in SEASONING of SOME AUSTRALIAN TIMBERS.

Original sizes of pieces, 6 in. x 4 in. x 4 ft. 6 in. Period of Seasoning, 3 years.

	Weight per cubic foot in 1887. LBS.	Weight per cubic foot in 1890. LBS.	Shrinkage per cent.	Loss of Weight per cent.
Red Ironbark	76.52	73 50	5.26	3.94
Grey Ironbark	73.85	72.64	4.85	1.64
White Ironbark	73.55	68.44	4.62	6.94
Tallow Wood	70.95	64.98	4.25	8.41
Black Butt	66.69	61.21	6.62	8.22
Spotted Gum	71.12	63.47	4.6	10.76
Red Gum	62.19	56.41	5.21	9.31
Flooded Gum	77.94	70.98	4.73	8.94
Grey Gum	57.33	45.3	3.69	20.98
Woolybutt	63.89	58.14	6.77	9.00
Mahogany	75.06	62.21	3.93	17.11
Forest Mahogany	72.23	62.23	5.86	13.08
Swamp Mahogany	75.98	61.6	9.9	18.92
Stringy Bark	71 33	57.24	6.72	19.75
Grey Box	73.62	67.73	8.53	8.0
Blackwood	70 58	45.5	2.55	35.5
Forest Oak	75.48	61.72	4.12	4.98
Turpentine	69.34	57.03	14.33	17.75
Pine	54.31	37.14	5.57	31.61
Rosewood	74.29	50 32	3.56	32.13
White Beech	63.03	52.87	2.29	15.96
Australian Teak	62 9	58.24	6.59	7.4

NEST DURABLE
COACH BODY
VARNISH.

NE HARD DRYING
COACH BODY
VARNISH.

BEST ELASTIC
OPAL CARRIAGE
VARNISH.

ST HARD DRYING
CARRIAGE
VARNISH.

ST QUICK DRYING
CARRIAGE
VARNISH.

Carriage Varnish

MANUFACTURED BY

Thomas Parsons & Sons

LONDON AND PARIS, ✳ ESTABLISHED 1802.

SOLD BY

FRANK GRIMLEY,

COACHBUILDERS' IRONMONGER,

263 and 265 CLARENCE STREET,

(NEAR THE TOWN HALL),

SYDNEY.

FINE
PALE WAGGON
VARNISH.

FINEST
BLACK
JAPAN.

BEST
BLACK
LACQUER.

FINE STRONG DRYING
JAPAN
GOLD SIZE.

BEST TEREBINE
OR
LIQUID DRYER.

peciality- BLACK JAPAN for Coachbody work.

Agent for Australasia- FRANK GRIMLEY,

CLARENCE STREET, SYDNEY.

Headquarters for Saddlery and Harness.

263 and 265 CLARENCE STREET,

NEAR THE TOWN HALL,

.....SYDNEY.

FRANK GRIMLEY

....IMPORTER and MANUFACTURER....

HARNESS — SADDLERY — LEATHER

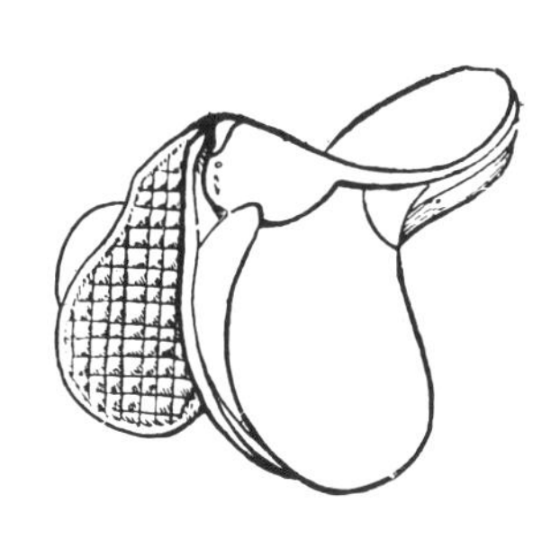

BICYCLES and BICYCLE PARTS.

BICYCLE MATERIAL. BICYCLE SUNDRIES.

SOLE DISTRIBUTING AGENT

..for..

"PERFECT" and "DOMINION"

...BICYCLES...

CARRIAGES AND THEIR PARTS.

Frank Grimley, Sydney.

JOHN RIGBY & SONS

ESTABLISHED 1842.

ADELAIDE 1887.

CELEBRATED

MELBOURNE 1888.

HIGH-CLASS AXLES,

AUSTRALIAN DRAY ARMS,
MAIL PATENT AXLES,
COLLINGE'S PATENT AXLES,
HALF-PATENT BUGGY AXLES,
SCREW NUT & TROLLY AXLES.
DRABBLE'S PATENT VAN ARMS,

.....ALSO BEST QUALITY.....

ELLIPTIC SPRINGS,
CONCORD BUGGY SPRINGS,
GRASSHOPPER & CART SPRINGS.
VAN & TROLLY SPRINGS.
BUGGY AXLE CLIPS, BOLTS ETC.,
COACH IRON WORK OF EVERY DESCRIPTION

WORKS & FORGE, WEDNESBURY, STAFFORDSHIRE

Highest Possible Awards obtained **SYDNEY 1879, MELBOURNE 1880-1881, ADELAIDE 1887, MELBOURNE 1888**

WRITE FOR PARTICULARS OF OUR NEW BUGGY WHEELS.

Showing FRANKLIN CONVERTIBLE BUGGY, Waggon Body.

SHOWN WITH TWO SEATS.

SARVEN WHEELS.

These wheels are put together on American principles. The bands are forced into position by hydraulic pressure (being the only way to make Sarven Wheels that will stand), making the most perfect Sarven Wheels that are manufactured in Australia.

Keep Bros. & Wood

IMPORTERS OF CARRIAGE BUILDERS' REQUISITES, MANUFACTURERS OF CARRIAGE BODIES & WHEELS

100 Franklin Street, Melbourne.

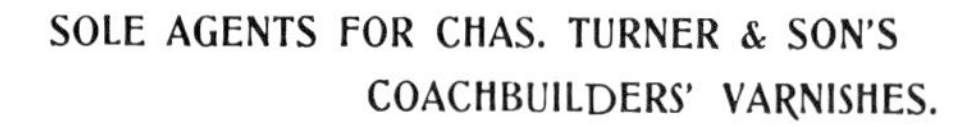

SOLE AGENTS FOR CHAS. TURNER & SON'S COACHBUILDERS' VARNISHES.

TRADE ONLY SUPPLIED.

Wheels Manufactured only from Guaranteed Dryest Timbers, and put together by Experienced Workmen. Write for Catalogues and Price Lists.

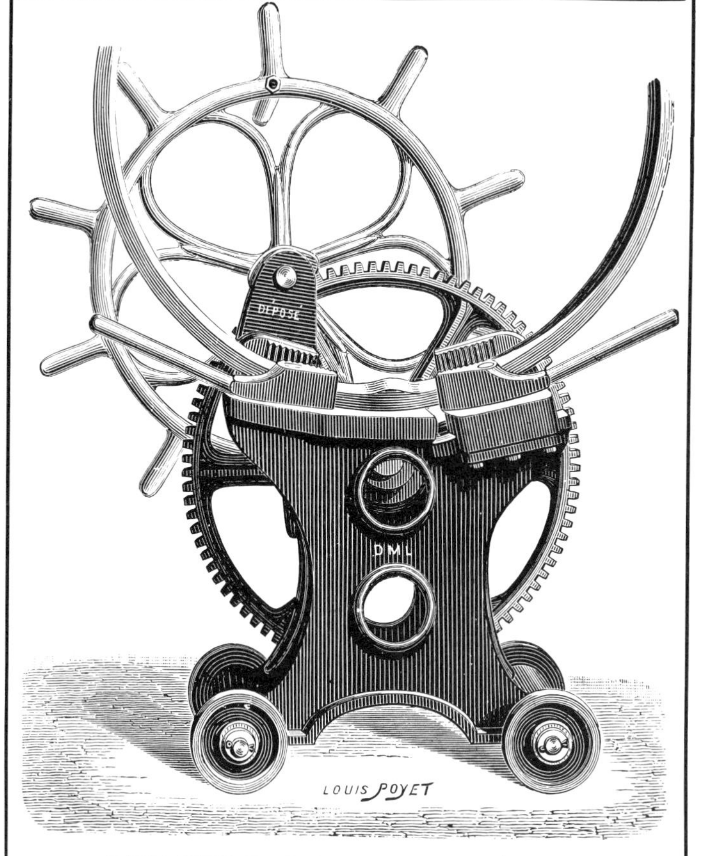

TYRE SHRINKING AND WELDING MACHINE.

In Heavy Sizes, 6in. x 1¼in. Medium, 4in. x 1in. Light, 3in. x 1in.

THE SHELDON AXLE COMPANY.

Anchor Brand Axles.

UNEQUALLED for FINISH and DURABILITY.

May be Obtained from all Coachbuilders' Ironmongers throughout the Colonies.

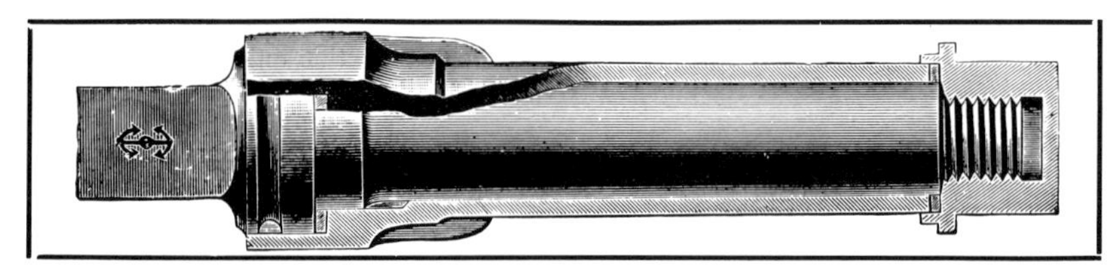

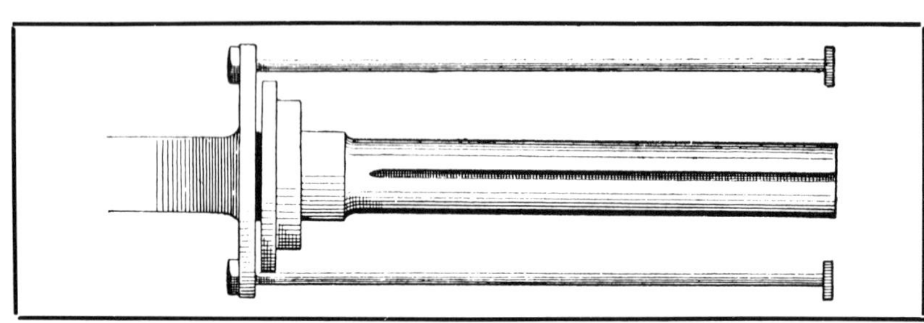

HALF-PATENT STEEL SINGLE OR DOUBLE COLLAR.

Square or round beds. Cast or wrought iron boxes.

English Pattern All Steel MAIL PATENT AXLE.

Beautifully Made.

RICHARDS LONG DISTANCE AXLE.

SINGLE OR DOUBLE NUT.

A boon to users of light or heavy carriages.

HALF PATENT STEEL AXLE.

Fitted with Collinge Collar.

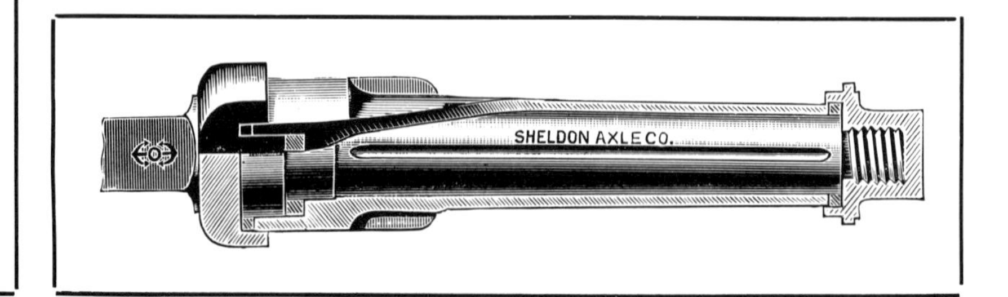

May be Obtained of All Hardware Merchants Throughout Australasia.

The Clifton Wheel Co.

Manufacturers of....

WHEELS, BODIES, and
Every Description of
CARRIAGE BUILDERS' WOODWARE

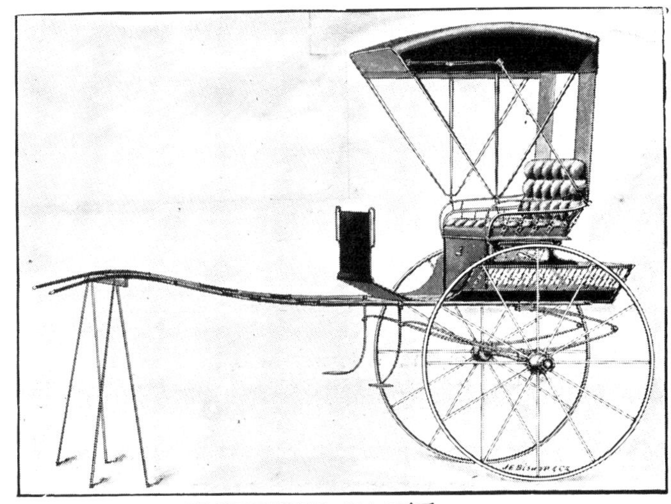

CLIFTON CART.

We hold the Patent Rights of Reid's Seat Shifter. ❧ ❧ Sole Manufacturers of the Lila and Clifton Patent Road Carts. Give us a Call. ❧ We are Up-to-Date.

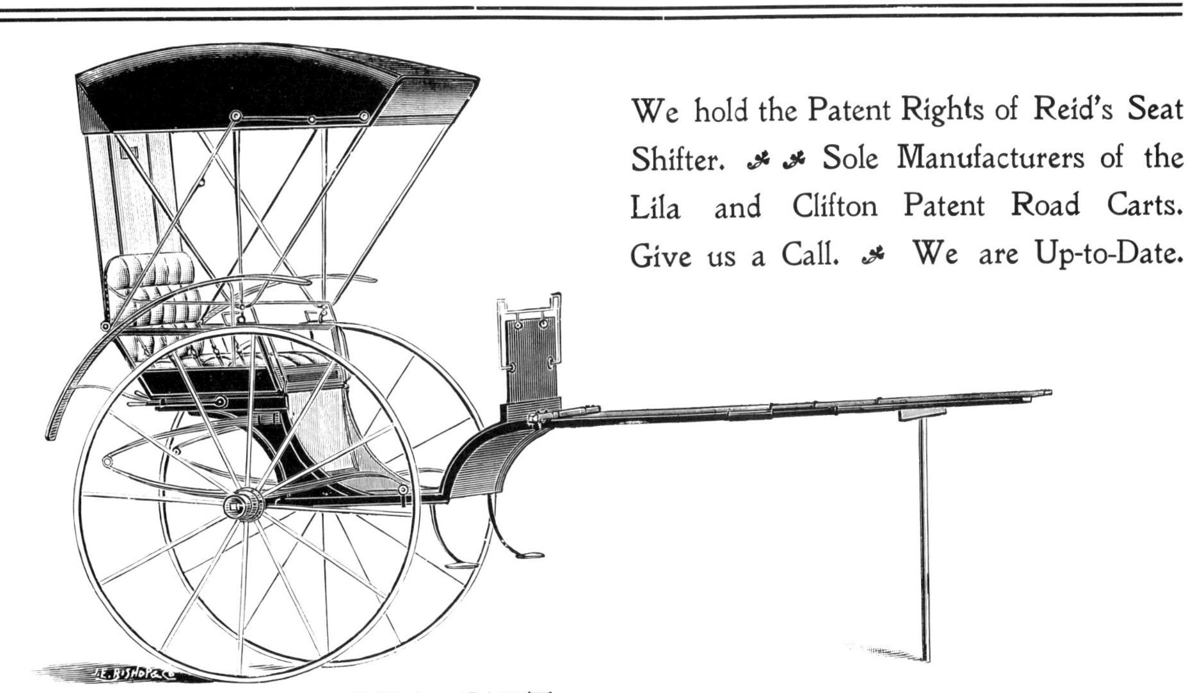

LILA CART.

62 Reilly Parade, Clifton Hill, Vic.

The "EASY" Perfect Rubber Tyres

FOR VEHICLES.

Supplied to the Trade only **Fitted with Wheels Complete.**

Leather=Covered Dashes and Plated Driving Rails.......

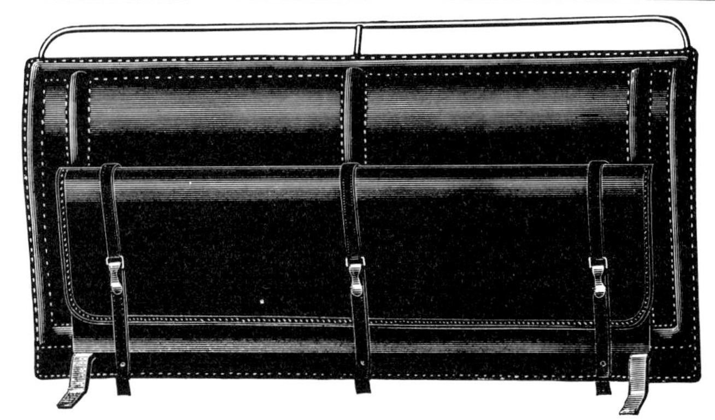

ABBOTT BUGGY DASH. 15in. high x 33 and 35in. long.

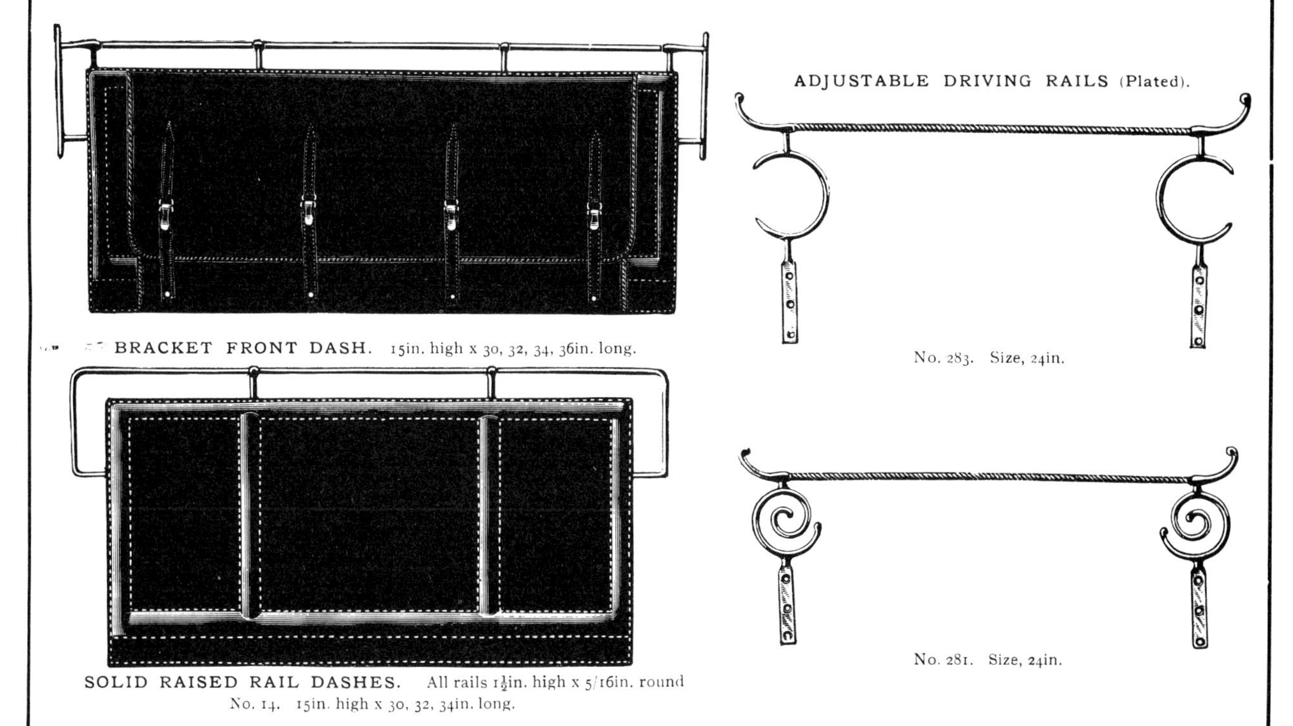

BRACKET FRONT DASH. 15in. high x 30, 32, 34, 36in. long.

ADJUSTABLE DRIVING RAILS (Plated).

No. 283. Size, 24in.

No. 281. Size, 24in.

SOLID RAISED RAIL DASHES. All rails 1¼in. high x 5/16in. round

No. 14. 15in. high x 30, 32, 34in. long.

KEEP BROS. & WOOD,

100 FRANKLIN STREET,
...MELBOURNE...

Do You Build Buggies?
CONCORD BUGGY.
Talking will not always sell a Vehicle.
OPEN ABBOTT BUGGY.
PONY PHAETON.
An Illustration will often make a sale.
Have Blocks of Buggies on your Stationery.
CURVED PANEL CART.
RUSTIC CART.
COAL BOX BUGGY.
BRASSEY CART.
STANHOPE ROAD CART.
An Electrotype
Of any of these Vehicles will be sent Post Free to any address in Australasia on receipt of 5s. 6d.
J. E. BISHOP & CO.
65 Market St., Sydney.
169 Queen St., Melbourne.
CURVED SIDE CART.
BAKER'S CART.
Blocks Copyright.
PHAETON SULKY.
SYDNEY SULKY.
A Block Catches the Eye, and Speaks forItself.....
Always Show a Block when advertising.
TURNOUT-SEAT BUGGY.
LINDSAY ROAD CART.
PIANO-BOX BUGGY.

ESTABLISHED 1791.
By Special Appointment to Her Majesty
THE QUEEN.
WM. HARLAND & SON'S
CELEBRATED
English Varnishes
THE WORLD'S STANDARD FOR QUALITY.
Used by all the leading railways, carriage and tram car builders and decorators. Unrivalled for brilliancy and durability.
Also
FINE COACH COLORS.
Head Office and Works: MERTON, LONDON, S.W., ENG.
Branches at Paris, Milan, Brussels, New York.
SYDNEY BRANCH: 82 PITT STREET, SYDNEY.
(WHOLESALE ONLY).
Melbourne Branch:
337 FLINDERS L., MELBOURNE.
VIRTVS
ET
LABOR